Gries / Debicki / Schind

Circular Textile Economy

Thomas Gries
Lukasz Debicki
Christian Schindler
Stefan Schlichter

Circular Textile Economy

HANSER

Print-ISBN: 978-1-56990-230-1
E-Book-ISBN: 978-1-56990-365-0
E-Pub-ISBN: 978-1-56990-532-6

Bibliographic information of the German National Library:
The German National Library lists this publication in the German National Bibliography; detailed bibliographic
data are available on the Internet at http://dnb.d-nb.de.

© 2026 Carl Hanser Verlag GmbH & Co. KG, Munich
Vilshofener Straße 10 | 81679 Munich | info@hanser.de
www.hanserpublications.com
www.hanser-fachbuch.de
Editor: Dr. Mark Smith
Production Management: Eberl & Koesel Studio, Kempten
Cover concept: Marc Müller-Bremer, www.rebranding.de, Munich
Cover design: Max Kostopoulos
Cover picture: © stock.adobe.com/Aija
Typesetting: le-tex publishing services GmbH, Leipzig

Contents

The Authors

The Editors

Univ.-Prof. Prof. h. c. (MGU) Dr.-Ing. Dipl.-Wirt. Ing. Thomas Gries

Thomas Gries, born in Cologne, Germany, in 1964, studied at the RWTH Aachen University, Germany. He holds a diploma degree in mechanical engineering and economics and a doctorate in mechanical engineering. From 1995 to 2001, he worked at Lurgi Zimmer AG, Frankfurt am Main, Germany, at the Department of Technologies for Fibres & Textiles in leading positions. Since 2001, he is Director of the Institut für Textiltechnik (ITA) of RWTH Aachen University. The honoris causa professorship of Lomonossow University, Moscow, is the most distinguished scientific award of Russia given for his achievement of 3D-tailored composite reinforcements. 2019 he was Distinguished Visiting Researcher at Seoul National University dedicated to 4D-Textiles. Today Thomas Gries is the thought-leader in 3D- & 4D-textiles. He made ITA the leading institution in digitalization and biotransformation of the textile sector.

Dr.-Ing. Lukasz Debicki

Lukasz Debicki studied Mechanical Engineering with focus on machine development and product design at RWTH Aachen University, Germany. In 2014 he joined the Institut für Textiltechnik Aachen (ITA) and has been working on the processes of melt spinning and texturing of polymeric multifilament yarns ever since. In his PhD thesis, he dealt with the topic of increasing the efficiency of melt spinning processes through digitalization approaches. Since 2020, Lukasz Debicki has been head of the Multifilament Technologies research group at ITA, whose research topics are in the areas of textile recycling, melt-spun elastic yarns from biopolymers, yarn functionalization and yarn structure formation.

Dr. Christian P. Schindler

Born in Karlsruhe, Germany, in 1968, Dr. Schindler studied economics at the University of Fribourg, Switzerland, from where he graduated in 1994. Between 1995 and 1998 he worked as personal assistant of two Members of Parliament in the German Bundestag, Bonn. In 1998 he joined the Federation of German Wholesale and Foreign Trage (BGA) in Bonn/Berlin as personal assistant and speechwriter of Association's President. Between 2001 and 2004 he studied at the Institute of Economic Policy at the University of Cologne, Germany, where he wrote his thesis and obtained a doctorate degree in 2004. Dr. Schindler was appointed Economist of the International Textile Manufacturers Federation on October 1, 2004 and was promoted to the position of Director in 2006. At the Federation's Annual Conference in Dubai, UA, in September 2006 he was nominated and elected Director General as of January 1, 2007.

Prof. Dr.-Ing. Stefan Schlichter

Stefan Schlichter graduated from RWTH Aachen University in 1982 in mechanical engineering. After an additional research period he specialized in high dynamic weaving and delivered a doctor thesis in 1987. In his professional career he specialized in different positions of the machine industry and worked in responsible positions in the field of research and development at Trützschler and later as a managing director in nonwoven machines at Oerlikon later Autefa Solutions. After nearly 30 years of experience in global machine business, he started to build up a new institute in Germany (ITA Augsburg) concentrating on composite recycling, artificial intelligence and textile recycling. In 2022 Prof. Schlichter established the "Recycling Atelier" as the worldwide first makers factory for mechanical recycling in cooperation between ITA Augsburg institute and the technical university of applied sciences in Augsburg. After retirement in the position of managing director of ITA Augsburg Prof. Schlichter concentrates on recycling and artificial intelligence in textiles as responsible manager for makers factories at technical university Augsburg.

The Contributors

Section	Contributors
1.1	Christian Möbitz, Thomas Gries
1.2	Henning Wilts, Burcu Gözet
1.3	Stefan Schlichter
2.1	Mauro Scalia, Felix Merkord
2.2	Fatah Naji, Wolfgang Rommel

Section	Contributors
2.3	Gabriella Waibel, Nicole Espey
2.4	Yan Yan
3.1	Uday Gill
3.2	Thomas Böschen
3.3	Priyanka Khanna, Laura Barbet
4.1	Stefan Schlichter, Amon Krichel
4.2	Annabelle Hutter
4.3	Hafiz Kaleem, Javier Vera Sorroche
4.4	Robin Sujatta, Matthias Schmitz
4.5	Lukasz Debicki; Stefan Schonauer; Ricarda Wissel
4.6	Edwin Keh
4.7	Sea-Hyun Lee, Stefan Schonauer, Sascha Schriever
5.1	Roxana Ley, Lukas Balon
5.2	Chiara Colombi, Erminia D'Itria
6.1	Vanessa Overhage
6.2	Mesut Cetin, Felix Teichmann, Georg Stegschuster

Preface to Circular Textile Economy

Global production is predominantly based on a linear model whereby new (virgin) resources are taken to produce products which are consumed and then disposed of by incineration or landfill. However, the concept of a circular model is not new. In segments such as paper card (around 71% in Europe in 2022), glass (around 80% in Europe in 2021), aluminum beverage cans (around 75% in Europe), or steel (around 55% in Europe) relatively high levels of recycling are already being achieved.

At the International Textile Manufacturers Federation (ITMF) Annual Conference 2022 which was held in Davos, Switzerland from September 18–20, 2022, the general conference theme was "Climate Change and a Sustainable Global Textile Value Chain". In the "Recycling Session," industry experts discussed recycling methodologies and technologies. In the traditional "Fiber Session" panellists shared their views on sustainable and circular fiber production. A presentation by EURATEX (European Textile & Apparel Industry Federation) highlighted how new legislation coming out of the Green Deal will affect the textile value chain. The presentations and the discussions among members illustrated that the topics of circularity, recycling and sustainability in the textile industry require a better and deeper understanding of the basic concepts of and the potentials for a circular textile economy.

Against this backdrop, the ITMF and the Institut für Textiltechnik Aachen (ITA) developed the idea to put together a webinar series with the title "Circular Textile Economy" during which a comprehensive overview would be provided about the legal framework, the fundamental concepts, the technologies and their applications.

In six webinars, various experts from the academia and the textile industry shared their knowledge and insights so that the participants could get a better understanding of the opportunities and challenges posed by a circular textile economy. Their presentations served as a basis for the articles in this book. ITA and ITMF would like to thank all the presenters/authors for their valuable contributions.

Against the background of contemporary policy objectives, the book systematically illuminates the entire process chain encompassing the manufacturing and recycling of textiles. It not only elucidates concepts for textile-to-textile recycling, but also delves deeper into these through industry case studies. Importantly, these concepts are extrapolated and, in the concluding chapters, translated into strategies for effecting a profound transformation towards Circular Economy principles.

Readers undertaking this journey through "Circular Textile Economy" will emerge with a profound understanding of the fundamental principles and practical implementations of textile-to-textile recycling. The book not only showcases existing industry solutions, but also unveils promising innovations that are poised to shape the future. As such, this work becomes the starting point for readers who seek to contribute tangibly to the preservation of our environment through concrete measures.

In the relentless expansion of global fiber production, the environmental toll exacted continues to escalate. To achieve the 1.5°C climate goal, a reduction in CO_2 emissions necessitates a paradigm shift through material substitution. While the proportion of recycled fibers is on the rise, much of it still comprises linear bottle recycling. The most promising avenue for a Circular Economy lies in textile-to-textile recycling, yet its current share in the pre- or post-consumer textile recycling market remains below 1%. As well as identifying these challenges, this book also presents viable solutions.

"Textile Circular Economy" aspires not only to inform, but to inspire a collective journey toward a more sustainable, circular future for the textile industry, driven by the collaborative efforts of esteemed editors and a network of global opinion makers and experts.

Thomas Gries, Stefan Schlichter,
Christian Schindler and Lukasz Debicki   Summer 2025

1

General Overview on the Topic: Textile Circular Economy

1.1 Circularity of Textiles – Threats, Limitations, Prospects

Christian Möbitz, Thomas Gries, Institut für Textiltechnik of RWTH Aachen University

1.1.1 Introduction

Sustainability and defossilization stand at the forefront of urgent global challenges. As we confront the scarcity of fossil fuels, the pressures of a burgeoning global population, and the profound impact of climate change, the textile industry is compelled to consider not only economic prosperity but also the long-term stewardship of our planet. The pivotal role of sustainable materials and production techniques, coupled with digital, integrative, and flexible production methodologies, cannot be overstated. Such strategies are essential in crafting individual business models that respond dynamically to these evolving challenges.

The textile industry's journey through time has been marked by significant transformations, particularly in the realm of material and energy sources. Historically, the industry relied on endemic and natural resources, a practice deeply rooted in the fabric of early textile production. However, with the advent of the industrial revolution, there was a monumental shift toward the utilization of wood and coal, and subsequently, the rise of petroleum as a dominant raw material source.

Today, the textile industry finds itself at a further critical juncture of another revolutionary era – the bioeconomy. This paradigm shift calls for a transition to more sustainable and regenerative resources, embracing biobased materials and closed-loop systems. This transformative movement is propelled by an acute awareness of the ecological impact of conventional textile production, coupled with advancements in materials science and recycling technologies.

Embracing the concept of bioeconomy is not just a strategic choice but a necessity. Bio-economy is an economic model that represents knowledge-driven utilization of earth's renewable resources, manifesting in innovative products, services, and procedures that permeate all economic sectors. It symbolizes growth that aligns with the rhythms of nature, striking a balance between development and environmental consciousness. Choosing bioeconomy ensures that the focus remains on both sustainability and economic feasibility.

As shown in Figure 1.1, 65% of the global fiber materials market today relies on petroleum-based products, underscoring the urgency to transition to more sustainable resources. On the renewables side, the environmental impact of cotton is significant, and the rising demand for cellulosic fibers and drop-in polymers highlights the need for rapid adoption of sustainable alternatives. A comparison of various fiber types produced from 1930 to 2030 reveals that synthetic fibers dominate, while the share of renewable resources has declined from 70% in 1972 to 30% in 2022. This trend further accentuates the need for sustainable practices in the textile industry.

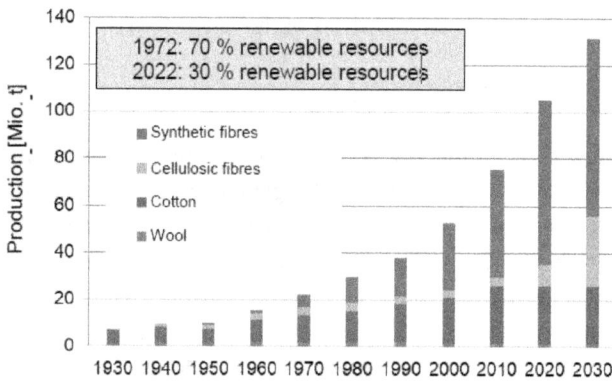

Figure 1.1 Development of sources for fibers [1]

The challenges highlighted by NIST [2], including the infrastructure for textile waste collection, sorting, grading, and commercial recycling processes, are critical aspects that require attention in the context of the textile circular economy. Furthermore, UNCTAD's [3] mention of the social dimensions, such as job creation and the management of end-of-life garments, underlines the necessity for a comprehensive, systemic and collaborative approach to enhance sustainability and circularity in the textile sector.

At this pivotal point of change, sustainability and digitalization are not merely trends but essential components for the textile circular economy and long-term success. Sustainable materials and manufacturing methods are vital for mitigating environmental impacts while meeting the needs of a growing global population [4]. Digital transformation, identified as a key driver of productivity enhancement in textile manufacturing, involves leveraging Industry 4.0 technologies, treating data as assets, and fostering change through leadership [5]. Tailored business models and strategies, attuned to specific customer needs, are crucial for success in a rapidly evolving industry. The effective use of data to address customer needs and offer customized solutions is increasingly becoming a competitive edge for textile companies [6].

In the quest for a textile-based bioeconomy, approaches like the renewable carbon strategy play a crucial role [7]. This initiative explores novel fiber sources, including CO_2 capture, biomass, and recycled materials, each serving as a foundation for renewable carbon that is adaptable to various textile applications. The journey extends beyond sourcing; it encompasses a holistic approach involving evaluation, developing new business cases and, crucially, scaling these innovations for widespread adoption. Delving into new raw material sources marks the beginning of a journey that necessitates a deep understanding of market viability, return on investment, and the nuances of large-scale production.

The cornerstone of textile circularity is the principle of recycling. It not only enables the repurposing of valuable resources but also contributes to reducing environmental impact and promotes the development of sustainable production methods. In this regard, recycling represents not just a technical challenge but also a strategic opportunity to align economic viability with environmental consciousness. The next phase in the evolution of the textile industry, focusing on recycling and reuse of materials, is thus a key element in the journey toward a truly sustainable and circular textile economy.

1.1.2 Approaches to Textile Recycling

Textile recycling, an integral part of the circular economy, involves transforming old or discarded textiles into new materials and products. This practice not only conserves natural resources but also reduces environmental pollution and fosters sustainable fashion. A prime example of this approach is seen in Prato, Italy, where the art of the circular economy has been perfected for centuries. Prato itself is renowned as one of the largest industrial centers in Italy and the largest textile center in Europe. The textile district of Prato consists of approximately 7,000 companies operating in the fashion sector and generating around 2 billion Euros through exports.

In Prato, companies like Comistra and Manteco have mastered the art of regenerating wool by recycling old wool fabrics. The process begins with the meticulous manual selection of rags, sorted by color, fineness, and composition. Following the removal of non-fabric components, like buttons, zippers, and labels, the material undergoes mechanical processing to create high-quality fibers. These fibers are then spun into yarns and knitted/woven into fabrics, producing unique colors without the need for additional dyeing or chemicals.

Manteco, in particular, has developed a closed-loop system that maximizes the use of raw materials, water, energy, and chemicals, ensuring sustainable textile production.

The firm not only recycles surplus raw wool from the production of raw woolen fabrics but also reuses the production waste from recycled wool fabrics.

Practices in Prato demonstrate how historical expertise, combined with modern sustainability standards, can transform an industry to be both ecologically and economically sustainable. A crucial step in Prato's recycling process is the careful sorting of textiles, which minimizes environmental impact and drastically reduces water usage, a significant factor in wool textile production. Recycled fibers are then purchased by fashion and textile brands to create new clothing with minimal waste generation.

Comistra, a company with over a century's presence in Prato, employs this method to mechanically convert waste into regenerated wool and new recycled yarns and fabrics. The company limits the use of dyes and chemical aids, thereby contributing to sustainable fashion. Comistra is also a founding member of Gida, a consortium aimed at reusing wastewater produced by textile companies in Prato.

Following this discussion on textile recycling practices, it is crucial to understand the broader framework guiding waste management – the waste hierarchy. As outlined in the EU Waste Framework Directive (Directive 2008/98/EC) and updated by Directive (EU) 2018/851, the hierarchy places waste prevention at its apex, as shown in Figure 1.2. This reflects the goal of minimizing the negative impact of waste generation and management and thus enhancing resource efficiency [8].

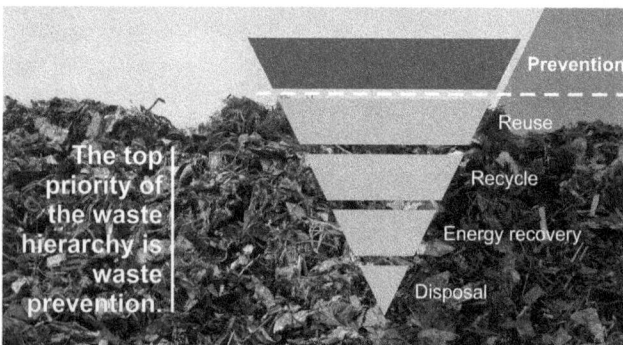

Figure 1.2 Waste hierarchy as outlined in the EU Waste Framework Directive

In Germany, this principle is embodied in the Circular Economy Act, which marks a paradigm shift from a disposal-based economy to a circular one. The objective is to conserve natural resources and manage waste in an environmentally friendly manner. The five-tier waste hierarchy includes prevention at the top, followed by reuse, recycling, other recovery (e.g., energy recovery) and, as a last resort, disposal [9].

EU member states are encouraged to adopt measures supporting sustainable production and consumption models and to promote the reusability and recyclability of products. Additionally, new recycling targets for municipal waste have been set, aiming for at least 55% by 2025 and increasing to 65% by 2035 [8]. In California, too, an extended producer responsibility program is under discussion [10]. In Germany, annual net waste generation amounts to approximately 350 million tons, with construction and demolition waste constituting the majority. The country has positioned itself as a leader in waste management and the circular economy, both nationally and internationally, through its advanced waste management system and the transfer of knowledge and technologies [11].

This approach to waste management underscores the significance of integrated strategies in textile recycling and the circular economy. By prioritizing waste prevention and efficient resource use, the textile industry can contribute significantly to achieving these ambitious recycling targets and fostering a more sustainable future.

The concept of *prevention*, as a top priority in waste hierarchy, is particularly relevant in addressing the challenges posed by the fashion industry. The issue of overproduction and excessive consumption in fashion has significant environmental and social implications. Fast-fashion, characterized by rapid production cycles and high raw material consumption, is a major contributor to environmental degradation. The swift changes in fashion trends lead to substantial textile waste, as clothing is often treated as disposable. Prioritizing materials of higher quality leads to more durable textiles and hence reduces the number of short-lived textiles.

The reuse and recycling options include the regeneration of fibers, both mechanically and chemically, the latter of which can produce high quality fibers similar to virgin fibers. In addition, there are processes such as anaerobic digestion, fermentation, composting, and thermal recovery that can help convert textile waste into other useful materials or thermal recovery to convert the waste into energy. Another approach is the conversion of textile waste into insulation or construction materials, which not only reduces waste but also reduces the demand for new materials and contributes to energy savings [12,13].

However, before reuse or recycling is even possible, it is necessary to collect the textiles. The textile recycling sector is developing rapidly and is of great interest to various stakeholders, due to its potential to reduce environmental impact. One study shows that interest in clothing made from recycled materials has risen from 57% in 2019 to 69% in 2020. However, the large volume of textile waste is also leading to an increasing breakdown in municipal textile collection, as collection rates for textile waste are often low and textiles often consist of mixed materials, a fact which makes sorting and pre-processing complicated and labor-intensive. Many recycling technologies are still in the development or pilot phase and have not yet reached commercial scale or profitability [13].

In conclusion, the journey through the various approaches of textile recycling reveals a compelling narrative of innovation and commitment to sustainability. From the historical expertise in Prato to the contemporary advances in recycling technologies and sustainable practices, the textile industry demonstrates a significant shift toward circular economy models. The focus on prevention, particularly in addressing the challenges of fast fashion, underscores the industry's goal of reducing its environmental impact and fostering social responsibility.

1.1.3 Fundamental Limitations of Recycling

Thomas Gries, Institut für Textiltechnik of RWTH Aachen University

A better understanding of the potential of textile recycling can be gained by studying the limitations of recycling, as they provide the scientific and technical framework. These limitations offer an insight into the complexity of, and potential for, recycling in the textile industry:

- Energy limitations
- Mechanical limitations
- Polymer-degradation limitations
- Functional limitations
- Economic limitations

These limitations are described in the following sections.

1.1.3.1 Energy Limitations

The energy limitations are based on the 2nd law of thermodynamics, which is a fundamental principle in understanding the behavior of systems in terms of energy and entropy. This law posits that the entropy of an isolated system either increases or remains constant over time, and highlights the inherent limitations in energy transformations.

In the context of recycling, this law is particularly pertinent. It underscores the energy limitations that must be considered when converting waste into useful products or energy. The principle suggests that the energy required to recycle materials should not exceed the energy recovered from the process. If it does, the recycling process will not be sustainable, as it will lead to a net increase in energy consumption and, consequently, entropy.

This limiation emphasizes the importance of developing efficient recycling methods that consume less energy and have a lower environmental impact. In practical terms, it means focusing on recycling processes that maximize material recovery while minimizing energy usage. This not only preserves the energy balance but also ensures that recycling is an environmentally viable and sustainable practice.

The application of this limitation in the textile industry involves evaluating the energy costs of recycling processes, from collection and sorting to the actual transformation of materials. By adhering to this principle, the industry can move toward more sustain-

able and responsible recycling practices, aligning with the broader goals of environmental conservation and resource efficiency.

Possible solutions

The challenge posed by the second law of thermodynamics to recycling can be addressed through sector coupling of renewable energies and recycling processes. The
illustration in Figure 1.3 shows how integrating various energy sources such as solar,
hydro, and wind power contributes to the generation of renewable hydrogen. This
hydrogen can then be combined with CO_2 from biomass to create renewable carbon,
which in turn is used in textile recycling [14].

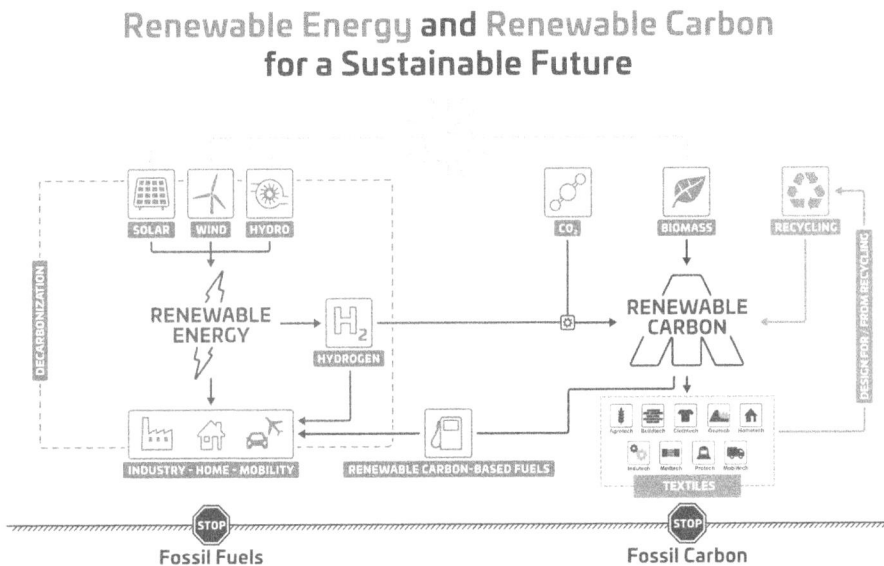

Figure 1.3 Sector coupling: A possible solution for the energy limits of recycling –
based on the Nova Institute's Renewable Carbon Initiative [14]

This coupling reduces dependence on fossil fuels and decreases CO_2 emission levels
typically associated with the recycling process. Using renewable carbon in textile recycling enhances sustainability by improving energy efficiency while simultaneously
minimizing environmental impacts. This model of sector coupling allows the overall
energy demand of recycling to be reduced by adapting to the thermodynamic limits.

1.1.3.2 Mechanical Limitations

Mechanical limitations emphasize a key challenge in textile fiber recycling. With each
cycle, the fibers become shorter. This phenomenon underscores an inherent difficulty
in the recycling of textile fibers. Recycling diminishes therefore the quality and reusability of the fibers. This degradation demonstrates that the physical properties of tex-

tile fibers can be compromised through repeated recycling. The mechanical strain of recycling weakens the fibers, posing a significant challenge for the sustainability of textile products.

The mechanical limitation presents a significant hurdle to achieving sustainable textile recycling, it necessitates the development of recycling methods that are less damaging to the fiber length and integrity.

Possible solutions

To overcome mechanical limitations, possible solutions are cascade applications, short-fiber separation and blending with longer fibers. The concept of cascade applications involves using fibers in a sequence of products with progressively lower requirements on fiber length. This approach effectively utilizes fibers throughout their life cycle, progressively downgrading their application as the fiber length diminishes. For example, fibers from high-quality lingerie can be reused in shirts, then jeans, followed by insulation materials, and finally in paper or cardboard. This sequence ensures that fibers are utilized in applications appropriate to their physical state, maximizes their use and reduces waste.

Short-fiber separation is a process whereby short fibers are segregated from longer ones, maintaining the quality of fibers for specific applications. This method is vital for ensuring that the recycled fibers meet the quality standards required for certain products. For example, extracting short fibers from business shirts allows these fibers to be specifically used in applications suited to short fibers and ensures that the quality of the material in high-value products is not compromised.

The third strategy involves blending shorter recycled fibers with longer, possibly virgin fibers, to increase the overall staple length. This method enhances the quality of the recycled material, making it suitable for higher-quality applications. Mixing shorter recycled fibers with longer new fibers maintains the overall integrity and strength of the textile material, allowing for its use in more demanding applications.

These strategies demonstrate that through creative and thoughtful processes, the challenges associated with fiber shortening in the recycling process can be overcome. They contribute to extending the lifespan of textile fibers, the efficient use of resources, and the promotion of sustainability in the textile industry. By adopting these methods, the industry can mitigate the effects of mechanical limitations and support the development of a more circular textile economy.

1.1.3.3 Polymer-degradation limitations

Polymer-degradation limitations in textile recycling addresses the challenge of decreasing molecular length of polymers with each recycling cycle. This reduction in molecular weight and crystalline fraction can lead to a deterioration of mechanical properties, such as decreased tensile strength and elasticity [15]. This underscores the need for developing recycling methods that minimize the impact on polymer molecular length and maintain the mechanical properties of recycled materials.

Possible solutions

To address the polymer-degradation limitations in textile recycling, several solutions are possible:

Solid-State Post-Condensation: This process aims to increase the molecular length of polymers after recycling. It involves further condensation of the polymer chains under heat and pressure in the solid state to yield longer polymer chains. This process can help restore some of the lost molecular length and improve the material's properties.

Partial Depolymerization and Repolymerization: Depolymerization involves breaking down the polymer chains of the recycled material partially, followed by repolymerization to achieve the desired polymer chain length. This prcesses can enhance the characteristics of the material for subsequent use, balancing the need for both recycling and maintaining material quality.

Complete Depolymerization and Repolymerization: This process entails fully breaking down the polymer chains into their monomeric components, followed by a new polymerization process to create new, long polymer chains. By starting from the monomers, this method allows for the production of polymers with tailored chain lengths and properties and effectively overcomes the limitations posed by the shortening of molecular chains during recycling.

Each of these solutions offers a way to mitigate the impact of recycling on polymer molecular length, thereby enhancing the quality and usability of recycled materials.

1.1.3.4 Functional Limitations

A functional limitation in textile recycling is the challenge given by mixtures of multiple materials. An example is the complex composition of a rain jacket, which often is a blend of materials. The jacket may feature polyester velvet with carbon particles and a polyester lining, both of which present recycling challenges, due to the difficulty of separation. Similarly, polyamide fabric, commonly used in such jackets, is also challenging to recycle, particularly when blended with other materials. Additionally, the presence of dyes and surface coatings, which often contain hard-to-remove, toxic chemicals, complicates the recycling process. Elements like polyoxymethylene and metal in zippers, as well as metal or plastic in buttons and logos, require removal before recycling. Moreover, components like polyester/polyamide hook-and-loop fasteners and adhesives, which are difficult to separate, can further hinder textile recycling. This complexity underlines the need for innovative separation and processing techniques to overcome these recycling challenges.

Possible solutions

The functional limitations presented by multi-material materials can be addressed with several strategies:

Design for Recycling: This emphasizes designing products with end-of-life recycling in mind. It involves using materials that are easy to separate and reuse and creating products that can be easily disassembled.

Predominantly Mono-Material Mix: Utilizing materials primarily consisting of one type simplifies the recycling process by reducing the complexity of material separation. Products made from single materials or compatible materials for recycling decrease the need for complex separation processes.

Comprehensive Life Cycle Assessment: Evaluating the entire life cycle of a product, from manufacturing to recycling, helps minimize environmental impacts and maximize material use efficiency.

These strategies highlight the importance of considering recyclability from the design phase and choosing materials that facilitate rather than hinder the recycling process. Implementing these approaches can improve material efficiency and reduce the environmental impact of products over their entire life cycle.

1.1.3.5 Economic Limitations

An economic limitation in circular economy emphasizes that transitioning to models of recycling often leads to increased costs. This is evident from the increased cost of collecting and sorting recyclable materials, especially when dealing with diverse or contaminated waste streams. The need for significant investments in new technologies and infrastructure for recycling and reusing materials also contributes to higher costs. Additionally, the manufacturing of products from recycled materials can be more expensive than using new materials, particularly if recycling processes lack optimization or if the quality of recycled materials is lower. These economic challenges underline the importance of developing efficient, cost-effective recycling, and circular economy strategies.

Possible solutions

These economic challenges may be addressed with the following innovative business models:

Digital Business Models and Producer Networks: These models leverage technology to create more efficient and sustainable production and distribution chains. Digital platforms can facilitate direct links between producers and consumers (B2C) or consumers and manufacturers (C2M), potentially bypassing traditional retail channels and reducing costs.

On-Shoring and Urban Production: Localizing production processes can result in higher wages, due to local labor costs, but also offers benefits such as reduced transportation costs and emissions. It also fosters innovation in sustainable practices through closer proximity to the consumption points.

These strategies highlight the need to balance initial investments with long-term sustainability goals. By implementing circular economy practices, industries can not only reduce their environmental footprint but also unlock new economic opportunities, drive innovation and create jobs. The shift to a circular economy model thus represents a transformative approach that balances economic, environmental, and social factors, which are essential for sustainable development.

1.1.4 Prospects on Textile Recycling

Textile recycling, particularly in the context of renewable energies and interdisciplinary collaboration, is evolving to address the environmental challenges of the textile industry. The integration of renewable energy sources like solar, hydro, and wind into CO_2-reduction and biomass-production processes is fundamental for enhancing the sustainability of the recycling process and transforming the textile industry into a more sustainable model.

Sustainable business models that incorporate social and environmental analyses, such as life cycle assessments (LCAs), are critical. These models help in understanding the overall impact of textiles throughout their life cycle, guiding the industry toward more sustainable practices. The transfer of laboratory innovations to industrial scaling is also essential for achieving substantial progress in textile recycling. This approach ensures that developments in textile technology and recycling methods are feasible on a larger scale, making the industry more sustainable.

The shift toward a more sustainable textile industry, guided by the principles of a circular economy, requires not just technological innovations but also the development of new forms of cooperation and partnership across various sectors and disciplines. This collaborative approach is necessary for addressing the complex challenges in textile recycling and fostering an environment where sustainable practices are the norm. It is about rethinking and redesigning the way textiles are produced, used, and recycled, and emphasizes the need for a systemic change toward sustainability.

Another approach that uses open innovation is the one adopted by *Recycling Atelier – ITA Augsburg (ita-augsburg.com)* which addresses high-level recycling instead of landfill and energy recovery. Within the Recycling Atelier, recycling is divided into several key steps:

- Design for Recycling: This approach focuses on creating textiles with recycling in mind, ensuring easier processing at the end of their life cycle.

- Processing and Preparation: This step involves preparing the textiles for recycling, which may include cleaning, disassembling, or other preparatory measures.

- Material Identification: Accurate identification of materials is crucial for determining the most appropriate recycling methods.

- Mechanical and Chemical Recycling: These are the core recycling processes, where textiles are either mechanically broken down or chemically processed to recover fibers or other materials.

- Upcycling: The final step involves transforming recycled materials into products of equal or higher value than the original.

The key to successful implementation lies in collaboration and joint innovation. There is a need to form cross-industry partnerships and advance the circular economy through combined efforts and the sharing of knowledge and resources. This integrated approach is essential for driving the textile industry toward a more sustainable and environmentally responsible future.

A third approach for further enhancing recycling product chains is the concept of the digital product passport (DPP). DPP is emerging as a transformative tool in the textile industry, particularly in fostering a sustainable and circular economy. The DPP, as part of the EU's Ecodesign for Sustainable Products Regulation, is designed to collect and disseminate comprehensive data about a product's life cycle, including its materials, supply chain, and environmental impact. This approach is expected to enhance transparency and accountability in the textile industry, enabling both consumers and manufacturers to access detailed information concerning a specific product [16, 17].

The digital product passport goes beyond traditional product labels by providing details such as the product's composition, origin, environmental footprint, repair, maintenance, and recycling instructions. It is linked to a unique product identifier and physically represented on the product with a data carrier such as a QR code. This accessibility allows consumers and stakeholders to easily access the information, empowering them to make more informed purchasing decisions and support brands that prioritize sustainability [16].

The implementation of DPPs is a response to the significant environmental impact of the textile industry, which is responsible for a considerable portion of global greenhouse gas emissions. By introducing DPPs, the EU Strategy for Sustainable Textiles aims to ensure that all textile products within the EU are durable, repairable, and recyclable, with a view to promoting transparency, sustainability, and responsible consumption in the apparel industry. In addition to benefiting consumers, DPPs also promote supply chain transparency. They enable stakeholders to identify sustainable practices and address potential issues; this facilitates responsible sourcing and production. Such transparency is crucial for building consumer trust and fostering a culture of accountability within the industry [16, 17].

The adoption of digital product passports by fashion brands is crucial for staying competitive in the EU market. Connected garments provided by DPPs not only offer transparency but also enable brands to personalize customer experiences, collect feedback, and adapt to changing consumer preferences. These passports, therefore, represent a significant step toward more sustainable and circular practices in the textile industry.

The implementation of DPPs, although promising, also comes with challenges. Standardizing data collection and recording methods across the industry and achieving industry-wide adoption require coordinated efforts among various stakeholders. Despite these challenges, the potential benefits of DPPs in driving sustainable change in the textile industry are substantial and represent a key component in the transition to a circular economy model.

In summary, the future of textile recycling lies in embracing renewable energies, fostering interdisciplinary collaborations, using digitalization technologies, and developing sustainable business models that prioritize environmental and social considerations. These approaches will lead to a more sustainable textile industry that not only meets the current needs but also ensures the protection and preservation of the environment for future generations.

1.1.5 Fields of Innovation for a New Circular Textile Bioeconomy

In pursuit of a circular textile bioeconomy, a number of possibilities and limitations have been discussed in the previous sections. In conclusion, a variety of key areas are crucial for the transition to more sustainable textile practices.

- **Fewer, Longer-Lasting, and More-Durable Products:** The industry aims to create textiles that not only last longer but also require fewer resources over their lifetime. By focusing on durability, we reduce the need for frequent replacements and, consequently, overall material waste.

- **Avoiding Overproduction:** Overproduction in the textile industry leads to surpluses that often go to waste. Innovative inventory and demand-prediction models, along with more responsive supply chain mechanisms, can help align production more closely with consumer demand.

- **Reuse and Repair Over Recycling:** Prioritizing the reuse and repair of textiles extends the life of garments and reduces the need to produce new materials. This can be achieved through services that facilitate repairs or by designing products with longevity and easy repairability in mind.

- **Textile Recycling as a Path to Sustainability:** Recycling textiles is essential for minimizing waste. Ongoing innovations are making textile recycling more efficient and less energy-intensive, turning post-consumer waste into valuable resources for new products.

- **Design for Recycling:** Incorporating recyclability into product design is fundamental. This includes using mono-materials where possible and designing for easy disassembly, thereby enabling more effective recycling processes.

- **Sustainable Distribution of Waste and Recycled Materials:** Optimizing the collection and distribution of waste and recycled materials ensures that they are returned to the production cycle efficiently. This reduces the carbon footprint associated with transportation.

- **(Political) Nudging:** Using policy tools to encourage stakeholders to make more sustainable choices, such as tax incentives for sustainable practices or penalties for wastefulness.

- **(Digital) Business Models:** Leveraging technology to create business models that facilitate a circular economy, such as platforms that connect consumers with recycling services or marketplaces for upcycled products.

- **Digital Product Passports:** Implementing digital passports that provide detailed information about a product's materials, origin, and end-of-life options to facilitate better recycling and informed consumer choices.

- **Purification Technologies:** Developing advanced methods to remove contaminants from recycled materials to ensure they meet quality standards for reuse.

- **Recovery Technologies:** Investing in technologies that can reclaim and repurpose materials from products that are no longer usable in their original form.

- **Disintegration Solutions:** Finding ways to break down materials into their original components, so that they can be made into new textiles.

- **Upcycling:** Creating higher-value products from waste materials; this not only diverts waste from landfills but also adds value to what would otherwise be discarded.

- **Repair Strategies:** Establishing systems and services that enable the repair of textiles, extending their useful life and reducing the need to produce new materials.

- **Biotransformation:** Exploring the use of biological processes, such as fermentation or enzymatic reactions, to transform waste materials into new textiles.

By focusing on these fields of innovation, the textile industry can continue to progress toward a more sustainable and circular future, reduce its environmental impact and lead the way for other industries to follow.

References for Section 1.1

[1]   Veit, D., Fasern – Geschichte, Erzeugung, Eigenschaften, Markt; doi: 10.1007/978-3-662-64469-0

[2]   Schumacher, K.A., Forster, A.L., *Frontiers in Sustainability*; doi: 10.3389/frsus.2022.1038323

[3]   Pacini, H., UNCTAD (2021), URL: *https://unctad.org/news/seizing-opportunities-circular-economy-textiles*

[4]   International Finance Corporation (2023), URL: *https://www.ifc.org/content/dam/ifc/doc/2023/strengthening-sustainability-in-the-textile-industry-ifc-2023.pdf*

[5]   INCIT (2023), URL: *https://incit.org/de/thought-leadership/digital-transformation-in-textile-manufacturing/*

[6]   Weisenberger, S., *IndustryWeek* (2017), URL: *https://www.industryweek.com/technology-and-iiot/article/22006120/sewing-digital-transformation-into-the-fabric-of-the-textiles-industry*

[7]   nova-Institut GmbH (2024), URL: *https://renewable-carbon.eu/*

[8]   EUR-Lex (2022), URL: *https://eur-lex.europa.eu/DE/legal-content/summary/eu-waste-management-law.html*

[9]   EUR-Lex (2023), URL: *https://eur-lex.europa.eu/DE/legal-content/glossary/waste-hierarchy.html*

[10] Gaetjens, B., *Recycling Today* (2023), URL: *https://www.recyclingtoday.com/news/textiles-california-sb707-epr-extended-producer-responsibility/*

[11] Umweltbundesamt (2023), URL: *https://www.umweltbundesamt.de/themen/abfall-ressourcen/abfall-kreislaufwirtschaft*

[12] Hussain, T., *The Textile Think Tank* (2023), URL: *https://thetextilethinktank.org/textile-waste-reuse-and-recycling-options/*

[13] Hussain, T., *The Textile Think Tank* (2023), URL: *https://thetextilethinktank.org/textile-recycling-latest-trends-challenges-and-opportunities/*

[14] Renewable Carbon (2024), URL: *https://renewable-carbon.eu/publications/product/renewable-energy-and-renewable-carbon-for-a-sustainable-future-%e2%88%92-graphic/*

[15] Semperger, O.V., Suplicz, A., *Sci. Rep.* (2023) 13, p. 17130, doi: 10.1038/s41598-023-44314-0

[16] Digital Link (2024), URL: *https://digital-link.com/product-passport/product-passport-for-textiles/*

[17] Protokol (2024), URL: *https://www.protokol.com/insights/digital-product-passports-in-the-textiles-industry/*

1.2 Ecological Rating Systems for a Circular Textile Economy

Henning Wilts, Burcu Gözet, Wuppertal Institut für Klima, Umwelt, Energie gGmbH

1.2.1 The Concept of a Circular Textile Economy

In recent decades, the demand for textiles and clothing has significantly increased, surpassing basic needs and heading toward excessive consumption [1]. This trend can be ascribed to an evolving textile industry that is ever more oriented toward mass consumption, commonly known as fast fashion. This swift expansion of the textile and apparel sector has brought about an intensified interconnectivity and complexity worldwide. As a result, the value chain for an item of clothing can span several countries and continents. For example, while the cotton for a T-shirt may be grown in Greece, spun into yarn in Turkey and processed into fabric in India, the stitching may be carried out in Bangladesh before the finished garment eventually finds its way to the European market [2].

The international competition among textile suppliers to establish a dominant position in this global market has propelled the relentless pursuit of cost optimization in value chains, resulting in the adoption of low-tech systems, the use of cheap materials, and the outsourcing of production processes to countries with low environmental and social standards and low labor costs. As a result, garments with significantly shorter lifespans have been and still are introduced that are increasingly regarded as disposable commodities within the market [3].

The European Union (EU) holds a pivotal position in the global textile market, surpassing the €200 billion milestone in textile and clothing trade for the first time in history in 2022. As the world's second-largest importer of textile, it imported 12.3 million tons of textiles and clothing in 2022 (+4.8% compared with the previous year), reaching an import value of €138 billion. The largest import partners are China, Bangladesh, and Turkey, with approximately half of the total import volume originating from China and Bangladesh. Simultaneously, the EU also forms the world's second-largest exporter of textiles and clothing, after China, with an export volume of 6 million tons in 2022 [4]. Hence, the EU positions itself as a significant trade partner and a crucial player, acting as both consumer and collaborative partner in manufacturing.

This underscores not only the EU's economic position and power within the textile industry but also its responsibility for the socio-ecological and environmental consequences along its entire supply chains. Alongside labor law violations, the negative impacts encompass the intensive use of raw materials and land, water consumption and contamination, as well as the release of greenhouse gas emissions and other pollutants into air and onto land [5]. Considering the EU's significant role as a major import partner, it is unsurprising that European consumption contributes substantially to those impacts. According to a study conducted by the European Environment Agency, textile consumption represents the second-largest impact category with regard to European land use, the fourth largest in terms of raw material and water use and the fifth largest in respect of greenhouse gas emissions – with the majority of these impacts taking place in extracting and producing countries outside the EU. This holds true for 85% of the primary raw materials use, 92% of the water consumption, 93% of the land use and 76% of the greenhouse gas emissions [6]. These widely interlinked complexities within the textile industry underline its need not only for transformation but also for a comprehensive approach in doing so.

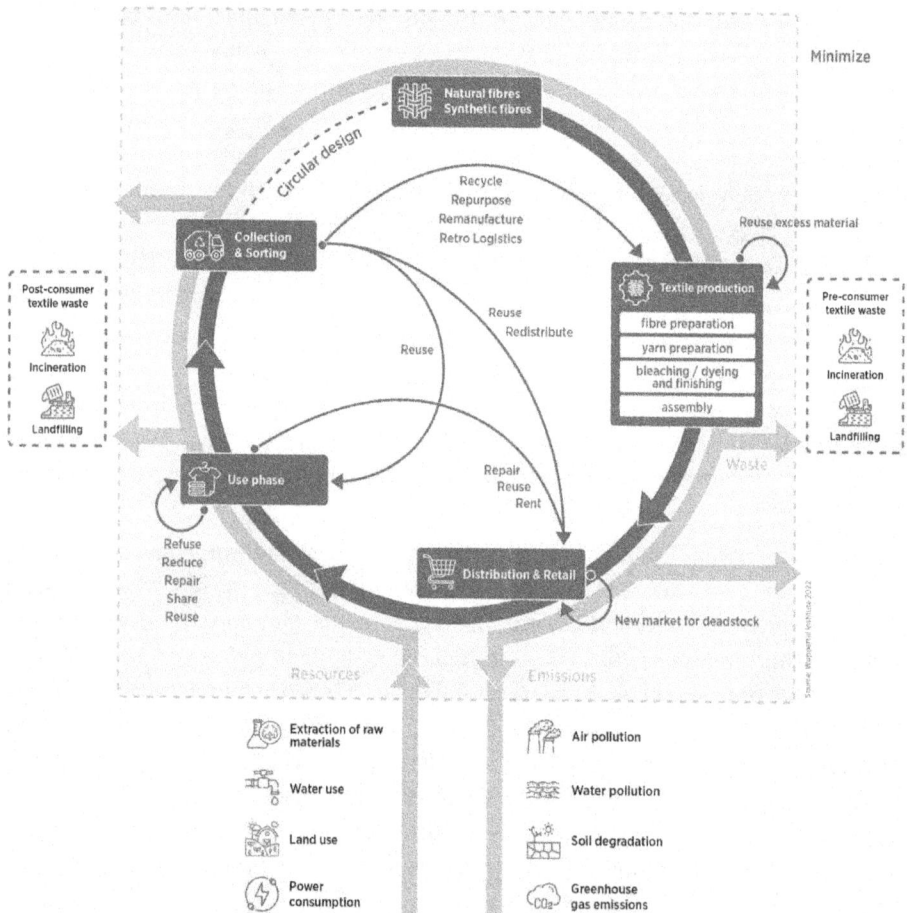

Figure 1.4 The concept of a circular textile economy [8]

The concept of a circular economy presents such a systemic approach and it can be realized by integrating circular strategies across the entire value chain. It is thus regarded as a source of leverage or a tool that contributes to reducing resource consumption and the associated negative environmental impacts [7]. Figure 1.4 illustrates how a circular textile economy of this nature can be envisioned at every stage of the value chain. It demonstrates that diverse circular strategies can effectively reduce both pre-consumer and post-consumer waste, consequently mitigating various negative environmental impacts throughout the textile system.

While circular measures are depicted for each stakeholder along the supply chain, businesses can play a pivotal role in implementing and enabling various circular resource loops. For instance, the design of a product can determine not only the product's lifespan but also the end-of-life treatment opportunities. Furthermore, circular business models can involve renting and leasing opportunities (i.e., product as a service), allowing customers to return their items and reduce the volume of new purchases.

1.2.2 Upcoming Political Steps and their Consequences for Businesses

As of 2022, in line with the European Green Deal, the Circular Economy Action Plan and the Industrial Strategy, the EU has released the EU Strategy for Sustainable and Circular Textiles. This strategy, with which the EU signals that it has taken on its responsibility as a key player in the global textile market, provides for the establishment of a circular textile market, where products are designed with a focus on durability, reusability, repairability, recyclability, and energy efficiency. This strategy encompasses initiatives throughout the value chain to enhance sustainable design and production processes and outlines a distinct vision for the textile industry by 2030:

> *"By 2030, textile products brought onto the EU market will be durable and recyclable, largely made from recycled fibers, free of hazardous substances and manufactured with respect for social rights and the environment. Consumers will also benefit from high-quality, affordable textiles. Fast fashion will no longer be fashionable, while economically profitable reuse and repair services will be widely available. In the competitive, resilient and innovative textiles sector, producers will take responsibility for their products along the value chain, especially when they become waste. This will be possible with the support of sufficient capacities for fiber-to-fiber recycling, which will minimize the incineration and landfilling of textiles" [9].*

Against this background, Table 1.1 outlines regulations, directives and initiatives initiated by the European Union that are either approved or in progress and are in line with the EU textile strategy's vision.

Table 1.1 Legislation and Their Key Contributions to a Circular Textile Economy [10]

Legislation	Key Contributions to Circular Textile Economy
EU Product Compliance Network	Harmonization of safety and labeling requirements for products, including chemicals and textile labeling standards.
Taxonomy Regulation Delegated Act	Incentivization of private investments in sustainable textiles
Eco-design for Sustainable Products Regulation (ESPR)	Design requirements for product durability, circularity, use of resources, reduce pollution
	Disclose product supply chain and environmental impact data with the digital product passport
	Accountability for textile waste with disclosure of number of discarded products
Empowering Consumers in the Green Transition	Defines what is considered greenwashing
	Restricted use of generic sustainability claims
Corporate Sustainability Due Diligence Directive (CS3D)	Mandatory risk analysis of entire supply chain
	Set preventive or remedial measures in prioritized areas of risk
	Annual due diligence reporting
EU Forced Labor Regulation	Prevent the sale of goods made with forced labor in the European Union
Green Claims Directive	Definition of the method on how to substantiate sustainability claims in an accessible way
PEF Methodology for Apparel and Footwear	Definition of a standard method to calculate and disclose product environmental footprint
Waste Framework Directive Extended to Textiles	Textile companies to financially contribute to the recycling infrastructures for textiles (EPR)
	Monetary benefit of producing circular products
Revision of Regulation on Waste Shipments	Stricter regulations on exports of textile waste to non-OECD countries
Revision of the Industrial Emission Directive	New norms for textile industry plants in the EU to reduce their environmental impact
Revision of EU Ecolabel Criteria	Alignment of the EU Ecolabel criteria with the eco-design requirements for textiles (ESPR)
Revision of the Textile Labeling Regulation	Definition of mandatory environmental data disclosures on the product label
EU Toolbox against Counterfeiting	Clarify roles and responsibilities of stakeholders to limit counterfeiting

This table not only shows the wide scope of activities and the profound transformation that will come about but also offers insights into the magnitude of the impact they will have on the stakeholders involved, including businesses. In the following, potential consequences for businesses will be outlined by way of example:

The development of a mandatory eco-design requirement [9] in order to improve the performance of textiles with regard to durability, reusability, repairability, and fiber-to-fiber recycling, for instance, may force companies to invest in new technologies and processes to meet those requirements. Further, textile businesses exporting to the EU may need to ensure that the products comply with the eco-design requirements to maintain access to the EU market.

Further, with the revision of the regulation on waste shipment [11], textile businesses will be obliged to demonstrate their commitment to environmentally friendly waste disposal practices, including recycling and reuse practices. For businesses operating in the EU, this may lead to expenses related to waste characterization, documentation, and transportation as well as to investments in sustainable waste-management practices.

With the ban on the destruction of unsold products [9], under the eco-design regulation, large companies will have to publicly disclose the number of products they discard, reasons for the discarded volumes and information on the volume of discarded products sent out for reuse, remanufacturing, recycling, energy recovery and disposal operations. For textile businesses, this will mean they will have to re-evaluate production and waste management strategies and potentially increase their investment in circularity measures.

Last but not least, on January 5th, 2023, the Corporate Sustainability Reporting Directive (CSRD) [12] came into force, establishing a more stringent and comprehensive reporting obligation on large companies located in the European Union. Companies have to provide a wide range of information and data points on sustainability issues in their annual management reports in order to make sustainability practices more informative and comparable. The requirements entail, among others [13]:

* Policies related to resource use and circular economy (disclosure requirement E5-1)

* Actions and resources related to resource use and circular economy (disclosure requirement E5-2)

* Targets related to resource use and circular economy (disclosure requirement E5-3)

* Resource inflows (disclosure requirement E5-4)

* Disclosure requirement E5-5 – Resource outflows (disclosure requirement E5-4)

Potential financial effect of resource use and circular economy-related impacts, risks and opportunities (disclosure requirement E5-6).

The challenges that come with the new regulations and especially with the Corporate Sustainability Reporting Directive (CSRD) can be manifold. One of them could be a lack of capacity and know-how within the company while another could be uncertainty about the starting point of the transition. However, the upcoming reforms can also be seen as an opportunity to gain consumers' trust and earn credibility. The CSRD can furthermore inspire businesses to innovate and to build a reputation for sustainability. But for this to happen, challenges need to be overcome for which access and use of certain tools may prove decisive. Ecological ratings can act as one such tool, serving to assess the performance of products, processes or the entire business with regard to its ecological impact.

1.2.3 Ecological Ratings as a Tool: Potentials and Barriers

Against this background of massive environmental challenges and at the same time fundamental regulatory changes, the actors along the textile value chain are currently seeking orientation and guidance for this transformation process. There is an abundance of initiatives, platforms and pilot projects – but ultimately transparent tools will be needed that provide an understanding as to which of these activities actually contribute to the overall objective of a circular textile industry as outlined above.

The current confusion in the market can easily be observed by looking at the broad range of labels and certifications that indicate a certain level of circularity or more broadly sustainability. According to the European Commission, more than 200 eco-labels are in use, of which more than half give vague, misleading or unfounded information, with 40% of those claims providing no supporting evidence and half of them offering weak or non-existent verification [14].

In March 2023, the European Commission put forward a proposal for the Green Claims Derective (Table 1.1) that would require companies to substantiate the voluntary green claims they make in business-to-consumer commercial practices by complying with a number of requirements regarding their assessment. In particular, they would have to take a full life cycle perspective and consider all significant environmental aspects and impacts when assessing and checking that a positive achievement has no harmful impacts on climate change, resource consumption and circularity, sustainable use and protection of water and marine resources, pollution, biodiversity, animal welfare and ecosystems [15].

Despite this high level of ambition, the proposed directive does not prescribe a single method for the assessment – companies and member states are left in charge of implementation, with the challenging task of developing their own understanding of environmental assessment approaches. In this context, ecological rating systems, such as life cycle assessments (LCAs), offer a way to come up with a consistent approach that fulfills the requirements outlined above and can also allow comparison of different solutions.

LCA is an established and widely recognized methodology for assessing environmental impacts associated with all the stages of the life cycle of a commercial product, process, or service – from raw material extraction and processing (cradle), through the product's manufacture, distribution and use, to the recycling or final disposal of the constituent materials (grave) [16]. A typical LCA study involves a comprehensive inventory of the energy and materials that are required along the supply chain and value chain of a product, process or service, and calculates the cumulative corresponding emissions to the environment. Owing to the complexity of products, relevant procedures for conducting LCAs have been developed and included in the 14000 series of environmental management standards issued by the International Organization for Standardization (ISO), in particular, ISO 14040 and ISO 14044. The last of these specifies requirements and provides guidelines for a life cycle assessment (LCA), including: definition of the goal and scope of the LCA, the life cycle inventory analysis (LCI) phase, the life cycle impact assessment (LCIA) phase, the life cycle interpretation phase, reporting and critical review of the LCA, limitations of the LCA, relationship between the LCA phases, and conditions for use of value choices and optional elements [17].

1.2.3.1 Limitations of Ecological Rating Systems

LCAs and alternative rating systems offer significant opportunities to demonstrate that a company, a product, or a process complies with given regulatory requirements. They can also provide very valuable insights for R&D (e.g., which potential product design might lead to lower environmental impacts). But despite these potentials, the limitations of the tool also need to be communicated transparently. A look at the characteristics of the textile sector shows that LCAs often aim to break down complex questions into single figures or easy answers. First of all, it should be clear that different life cycle assessments for the exact same product can lead to very different results depending, for example, on the definition of system boundaries and the selection of impact categories to be included in the analysis. The ISO standards offer a reference framework, but they do not precisely define how to conduct an LCA – thus emphasizing the importance of formulating the question that an LCA is supposed to answer and especially the careful interpretation of the results. Figure 1.5 highlights different life cycle models that all can be used to conduct an LCA. Cradle-to-gate focuses just on the production processes and is often used for environmental product declarations. Cradle-to-grave adds the use phase and disposal stage while cradle-to-cradle considers a closed loop recycling system. Obviously, the choice of life cycle model changes the outcome of the analysis and thus calls for a very careful consideration of the method selection that nevertheless rarely happens in reality [18].

LCAs are often criticized for the dependance of the outcomes on the design of the analysis. However, this is more a strength of the approach than a weakness. It allows complex problems to be looked at from different angles. This requires, of course, a clear understanding of the process on the part of those conducting and, especially, using LCAs.

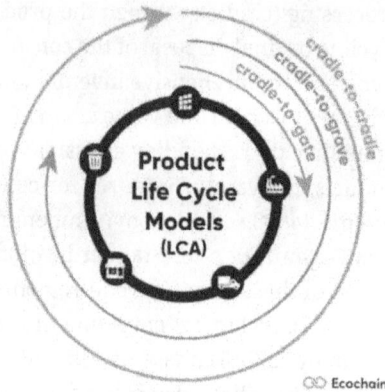

Figure 1.5

Life cycle models for an LCA [19]

One concrete example that highlights the need for a differentiated view of LCA results is the use of recycled plastic in the textile industry, especially for clothing. Companies communicate this as a way of reducing the carbon footprint of their products, rightfully referring to life cycle assessments which indicate that recycled PET causes significantly lower environmental impacts than virgin PET, according to the Joint Research Center of the European Commission [20]. This means that, at the product level, clothing with a higher content of recycled plastic has less impact in terms of, for example, greenhouse gas (GHG) emissions.

At the same time, a life cycle analysis could also consider the limited availability of high-quality recycled PET (rPET). As a result of insufficient collection, sorting and recycling infrastructures, the demand for rPET clearly exceeds supply. This means that a different perspective could include alternative uses for the material in textiles or, for example, in plastic bottles. Compared with textile waste, plastic bottles have a much higher separate collection rate and will be recycled with much higher probability. From such a more comprehensive point of view, an LCA could indicate that the use of rPET in clothing leads to higher GHG emissions because it means that more virgin PET will have to be used for plastic bottles.

1.2.3.2 Conclusions and Future Need for Research

As described in the previous sections, the textile industry, including the entire value chain, currently faces the need for a fundamental transformation. The level of environmental impact caused by the production and consumption of textiles is clearly not compatible with the need for future climate neutrality and for a significant reduction in resource consumption on a global scale. Over recent years, policy makers have switched from information-based approaches, which appealed more to the environmental consciousness of companies and consumers, to much more stringent policy tools and market interventions.

Against this background, the use of ecological rating systems will gain in importance because either it will become mandatory or relevant stakeholders will ask for science-based proof of improved environmental performance of products. Here, especially, the EU taxonomy could become an important driver for ecological rating systems as they might become an important prerequisite for access to financing and funding.

Examples such as that concerning rPET in clothing do not cast doubt on the potential contributions of life cycle assessments. They simply highlight the need for the careful use of such ecological rating systems. Formulating meaningful questions that then can be answered by LCAs requires an in-depth understanding of the current system, changes in the regulatory framework, and technology developments. Focusing on closed loop or circular solutions, especially, needs a detailed understanding of consumer behavior. How long will they use a certain product? How do they dispose them at the end of the use phase? And what kind of rebound effects would be generated if circular solutions were actually to lead to relevant cost savings?

Given this background, there will be a need for more interdisciplinary research focusing on the interlinkages between technology changes and resulting changes of consumer behavior. The "attitude behavior gap", for instance, highlights that the potential environmental benefits of a product are often not realized in practice: the potential recyclability of products and components is an important step but can even lead to increased resource consumption if, in the end, textile waste is not properly collected and recycled.

A second obvious key challenge is the quality of available data on which life cycle assessments are based. The complexity of textile value chains that typically span several continents creates high uncertainty about how data on environmental impacts have been generated. It is also often unclear to what extent assumptions about, for example, recycling routes actually reflect realities or have been based on outdated or simply wrong data. There is intense debate about how the quality of data used for LCAs might be assessed in order to allow a better understanding of the accuracy of results. DIN, the German Institute for Standardization, has coordinated a standardization roadmap for a circular economy that aimed to highlight current gaps and future need for additional or updated standards. Here especially, a need for clear metrics on data qualities used in LCAs have been mentioned but so far without a clear idea of how to address this challenge [21].

In order to increase the transparency of value chains, the European Commission has initiated the establishment of a digitally supported circular economy, based inter alia on digital product passports (DPPs) and a smart data space for the circular economy. Despite it still being in its infancy, DPPs especially could become a game changer for life cycle assessments, too, if they allow the use of real-time data instead of generic or average data [22]. But of course, this will also open up various research questions: what kind of data would be necessary for such assessments, how could they be gathered and especially who would have access to that kind of highly sensitive data? The following chapters of this publication will hopefully provide new insights that will help to answer these questions.

References for Section 1.2

[1] Industrievereinigung Chemiefaser e.V. (IVC), *Die Chemiefaserindustrie in der Bundesrepublik Deutsch-land 2020/2021* (2021), *https://www.ivc ev.de/sites/default/files/informationsmaterial dateien/IVC%20 Jahresbrosch%C3%BCre%202021.pdf*

[2] Köhler, A., Watson, D., Trzepacz, S., Löw, C., Liu, R., Danneck, J., Konstantas, A., Donatello, S., & Faraca, G., *Circular economy perspectives in the EU textile sector*, Publications Office of the European Union (2021), *https://data.europa.eu/doi/10.2760/858144*

[3] Remy, N., Speelman, E., & Swartz, S., *Style that's sustainable: A new fast-fashion formula*, McKinsey Sustainability (2016). *https://www.mckinsey.com/business functions/sustainability/our-insights/style-thats-sustainable-a-new-fast-fashion-formula*

[4] Euratex, Spring report: Analysis of the EU textile and clothing external trade in 2022, Brussels: The European Apparel and Textile Confederation, (2023)

[5] Europäische Kommission, *Data on the EU textile ecosystem and its competitiveness: Final report*, Publi-cations Office of the European Union (2021), *https://data.europa.eu/doi/10.2873/23948*

[6] Manshoven, S., Maarten, C., Vercalsteren, A., Arnold, M., Nicolau, M., Lafond, E., Fogh Mortensen, L., & Coscieme, L., *Textiles and the environment in a circular economy,* No. 6/2019; European Topic Centre on Waste and Materials in a Green Economy, (2019). *https://www.eionet.europa.eu/etcs/etc-wmge/ products/etc-wmge-reports/textiles-and-the-environment-in-a-circular-economy*

[7] Gözet, B., & Wilts, H., *Transforming our World« – Zukunftsdiskurse zur Umsetzung der UN-Agenda 2030,* Kreislaufwirtschaft als Baustein nachhaltiger Entwicklung. In C. Meyer (publisher) (2022) pp. 173–180. *https://www.transcript-verlag.de/978-3-8376- 5557-5/transforming-our-world-zukunfts diskurse-zur-umsetzung-der-un-agenda-2030/*

[8] Gözet, B., & Wilts, H., *The circular economy as a new narrative for the textile industry: An analysis of the textile value chain with a focus on Germany's transformation to a circular economy* (Zukunftsimpuls no. 23). Wuppertal Institute. (2022). *https://epub.wupperinst.org/frontdoor/deliver/index/docId/8108/ file/ZI23_Textile_Industry.pdf*

[9] Directorate-General for Environment, *EU Strategy for Sustainable and Circular Textiles.* COM 141 final Brussels: European Commission, (2022). *https://ec.europa.eu/environment/publications/textiles strategy_en*

[10] Trustrace, *Cheatsheet for The EU Sustainable Textile Strategy*, SWIN Technologies AB. (2023) *https:// trustrace.com/downloads/eu-sustainable-textile-strategy-cheatsheet*

[11] Directorate-General for Environment, *Proposal for a new regulation on waste shipments,* COM 709 final. Brussels: European Commission, (2021). *https://eur-lex.europa.eu/legal-content/EN/TXT/?uri= CELEX%3A52021PC0709&qid=1642757230360*

[12] European Commission, *Proposal for a Directive of the European Parliament and of the Council amending Directive 2013/34/EU, Directive 2004/109/EC, Directive 2006/43/EC and Regulation (EU) No 537/2014, as regards corporate sustainability reporting,* COM 189 final, (2021). *https://eur-lex.europa.eu/legal-content/ EN/TXT/?uri=CELEX:32022L2464*

[13] EFRAG, *ESRS 5 Resource use and circular economy,* European sustainability reporting standards, (2022). *https://www.efrag.org/Assets/Download?assetUrl=%2Fsites%2Fwebpublishing%2FSiteAssets%2F12%2520 Draft%2520ESRS%2520E5%2520Resource%2520use%2520and%2520circular%2520economy.pdf*

[14] European Commission, *Green Claims, New criteria to stop companies from making misleading claims about environmental merits of their products and services,* (2023)

https://environment.ec.europa.eu/topics/circular-economy/green-claims_en

[15] Ragonnaud, G., *'Green claims' directive, Protecting consumers from greenwashing*, European Parliamen-tary Research Service, (2023). *https://www.europarl.europa.eu/RegData/etudes/BRIE/2023/753958/EPRS_ BRI(2023)753958_EN.pdf*

[16] Umwelt Bundesamt (UBE), *Ökobilanz*, (2018). *https://www.umweltbundesamt.de/themen/wirtschaft-konsum/produkte/oekobilanz*

[17] International Organization for Standardization, *Environmental Management – Life cycle assessment: Requirements and Guidelines*, ISO standard no. 14044:2006 (2006) *https://www.iso.org/standard/38498. html*

[18] Bitter-Krahe, J., *Situation-oriented approach selection for sustainability assessments*, (2021) Aachen Technische Hochschule, Doctoral thesis *https://epub.wupperinst.org/frontdoor/index/index/start/5/ rows/10/sortfield/year_sort/sortorder/desc/searchtype/simple/query/krahe/docId/8012*

[19] Ecochain, *Life Cycle Assessment (LCA) – Complete Beginner's Guide*, (2023). *https://ecochain.com/blog/ life-cycle-assessment-lca-guide/*

[20] Nessi S., Sinkko T., Bulgheroni C., Garcia-Gutierrez P., Giuntoli J., Konti A., SanyeMengual E., Tonini D., Pant R., Marelli L., Comparative Life Cycle Assessment (LCA) of Alternative Feedstock for Plastics Production – Part 2, JRC Technical Reports. European Commission, Ispra, (2020)

[21] DIN, *Standardization Roadmap Circular Economy*, (2023). *https://www.din.de/en/innovation-and-research/ circular-economy/standardization-roadmap-circular-economy*

[22] Jansen, M., Gerstenberger, B., Bitter-Krahe, J., Berg, H., Sebestyén, J., Schneider, J. *Current approaches to the Digital Product Passport for a Circular Economy* (Wuppertal Paper no. 198). Wuppertal Institute, (2022). *https://epub.wupperinst.org/frontdoor/deliver/index/docId/8042/file/WP198.pdf*

1.3 Textile Recycling – Tools for Textile Economy

Stefan Schlichter, Technische Hochschule Augsburg (University of Applied Sciences)

1.3.1 Current Situation of the Textile Cycle

Textile production is currently characterized by a predominantly linear process chain. As a result, the cycle of textile value creation is not considered in its totality, as only a linear sequence from production through use to disposal is applied. As a consequence of this approach, the following typical phenomena can be observed, which both highlight the weaknesses of linear production and describe the deficits that should be avoided when implementing a circular textile production:

- 73% of used textiles are either subjected to energy (thermal) disposal or are deposited [1].

- The high diversity of materials commonly used in textile production poses challenges in recycling, due to a lack of options for material separation.

- Global textile consumption continues to grow unabated, due to trends like fast or ultrafast fashion, leading to large quantities of mostly low-quality items where reuse or recycling is economically unfeasible.

- The collection of used textiles mainly focuses on clothing textiles, while technical textiles (e.g., automotive, filtration, geotextiles) or textiles from the hygiene, home textile, or construction textile sectors are practically not collected and therefore not recycled [2].

- Even in countries that have high collection rates for clothing textiles (e.g., Germany at 1.6 million tons/year [1]), only a small portion of the collected and sorted textiles are reused or recycled.

- The recycling processes employed for reutilization primarily result in lower quality (downcycling) and end up in applications such as cleaning cloths, painter's cover fleeces, or car interiors. These are products that do not allow for high value creation through the use of secondary raw materials.

■ As a consequence of the facts mentioned here, only 1% of used textiles in the European Union are recycled for renewed use of the fibers in textile products (fiber-to-fiber recycling).

■ Future political directives (e.g., EU textile strategy [4]) will mandate the textile industry to aim for the complete realization of a circular textile value chain.

■ Consumer behavior, especially among younger consumers, will favor sustainable textile products in the future.

As a vision for a modern textile cycle, a new concept for implementing a value chain must be developed that considers the complex interrelationships described above and gives more prominence to user behavior as well as individual aspects of textile process chains. The goal should be to obtain higher qualities from recycling strategies, thereby avoiding thermal disposal and landfilling. The ideal textile cycle consists not only of a single cycle but a circular arrangement of multiple cycles that considers different material compositions and qualities and represents a spiral of different usage forms as in the example of the textile recycling options proposed in Figure 1.6 [3].

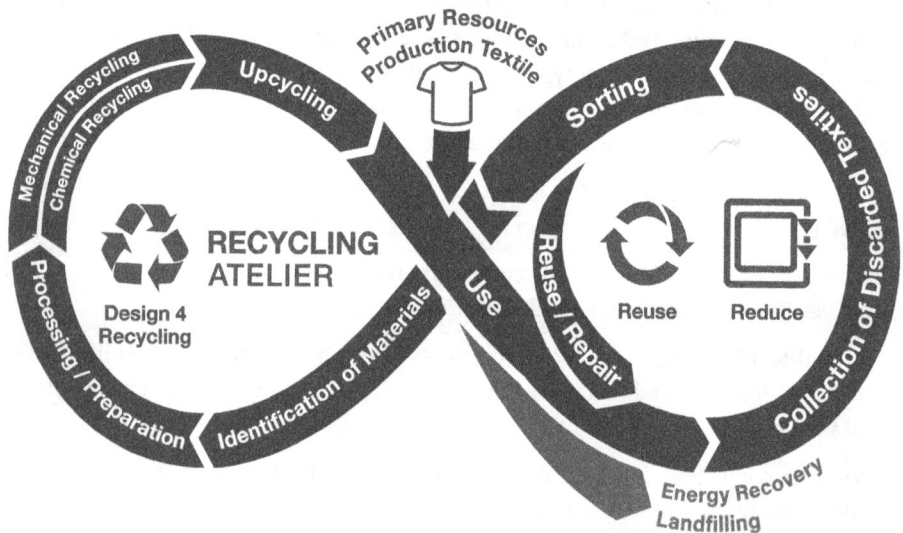

Figure 1.6 Vision of a modern textile cycle to avoid thermal utilization and landfilling of textiles (realized in the Recycling Atelier's model factory)

However, a future textile circular economy must ensure all previous (technical and economic) requirements for a textile product are met while also meeting additional new (ecological, waste-legislation-related and social) standards. Future competitiveness must focus on these new aspects. These conditions indicate that product development for re-cyclable textiles has become much more complex, requiring new tools that enable tar-geted development of new and, accordingly, more competitive products that meet the specified requirements.

1.3.2 Need for New Tools

The diversity of materials used in textile manufacturing described at the outset and the complex textile value chain pose new challenges that can only be overcome with new tools for the development of high-quality textiles from secondary materials. The entire value cycle of a future circular economy must be considered in every individual sec-tion of the process chain. The various requirements of the product development pro-cess must be adapted in each value-added step and their impact on the desired textile cycle and the requirements of product development must be taken into account. The following example describes a concept for addressing these new requirements in each process step, from collected used textiles to the new textile product.

The example concerns a model factory for the development of new textile products from secondary raw materials, as realized in the form of the Recycling Atelier Augs-burg. The need for a model factory arises from the fact that the complete value chain within product development cannot be technically represented without incurring un-reasonably high expenses in terms of space and costs. A practical way to represent the required short development cycles must therefore enable process-compatible imple-mentation of the value chain with significantly smaller amounts of material.

The model factory must be able to equally represent various requirements, the key ones of which are listed below:

* Materials
* Process
* Technologies
* Logistics
* Design

1.3.3 Concept of a Modern Toolbox for Recycling

The concept of the Recycling Atelier model factory, as shown in Figure 1.7, illustrates all the stages of value creation, from used textiles to new textile products, in the form of model-like individual steps. The configuration of the Recycling Atelier follows the mate-rial flow and includes individual stations that correspond to each value-added step in the textile processing chain. The concept of the model factory ensures that even small quantities of textiles can be used for product development by modeling each step. Transfer to the necessary industrial scale-up in the subsequent stages must be ensured at all times. All individual steps of the model factory also adhere to the concept of so-called upcycling, meaning a focus on high-level quality in each individual process step. Through a partner concept, industrial partners from the fields of process technology, mechanical engineering, quality assurance, and material production are integrated to support each process step. This ensures that future upscaling to industrial production conditions can be implemented at any time.

1. Material analysis
2. Sorting
3. Preparation   5. Spinning mill
4. Textile   6. Product design
   processing   7. Workshop

Figure 1.7 The Recycling Atelier cycles textile secondary raw materials back into high-quality products

1.3.4 Tool Implementation Using the Example of the Recycling Atelier

The implementation of new tools that meet the described requirements will be exemplified below, based on selected practices of the Recycling Atelier. However, in principle, these concepts can also be applied to other development concepts and strategies that aim for the holistic implementation of a textile circular economy.

1.3.4.1 Sorting

The sorting of collected used textiles plays a pivotal role in ensuring high-quality standards for the subsequent recycling process. In industrial practice, manual sorting is still commonly employed, incorporating various sorting criteria based on the desired reuse concept or recycling strategy. However, manual sorting has several weaknesses that necessitate the consideration of new concepts in this area. Firstly, manual sorting is cost-intensive, prohibiting economically viable applications for the future. Additionally, the limited sorting capabilities of manual inspection compromise the quality of sorting, making it unsuitable for meeting the requirements of high-value further textile utilization.

Therefore, the Recycling Atelier is actively exploring concepts for automatic sorting, even on a model scale, investigating different sensors and tailored software for intelligent and high-quality sorting in the future. In this regard, the advantage of a model factory becomes apparent in the implementation of new concepts, sensors, and intelligent evaluation strategies, which can be conceptually and experimentally examined and developed before deployment in the industry.

Figure 1.8 juxtaposes a typical workstation for manual sorting on the left and, on the right, a test setup for automatic sorting that allows the use of various sensors and evaluation strategies.

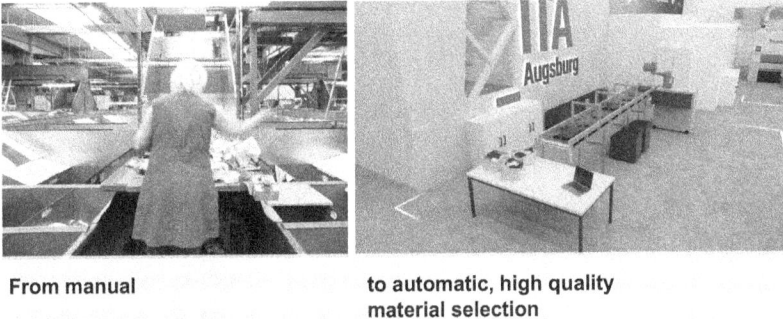

From manual

to automatic, high quality material selection

Figure 1.8 Sorting – from manual to automatic, high quality material selection (copyright: TexAid and ITA Augsburg)

1.3.4.2 Process Selection

The textile value chain is highly diverse and features combinations of process stages for the realization of tailored new products. It is therefore essential for the model factory to be able to represent these process variants, at least at a fundamental level, so that the right process for manufacturing a high-quality new textile product can be captured. Given the multitude of different procedural steps, it is not feasible to represent all textile process variants. However, the model factory must ensure that fundamentally important process variants can be tested on a model scale in order that the correct process chain for the desired end product may be created.

As depicted in Figure 1.9, the model factory can accommodate both yarn-based processes in spinning and fiber-based nonwoven fabric production. In addition to the assessment of different concepts for textile area production, this strategy has the advantage of allowing processing of partial and residual quantities from the utilization of textile secondary raw materials. This is indispensable in view of the comparatively high amounts of waste generated in the processing of recycled materials.

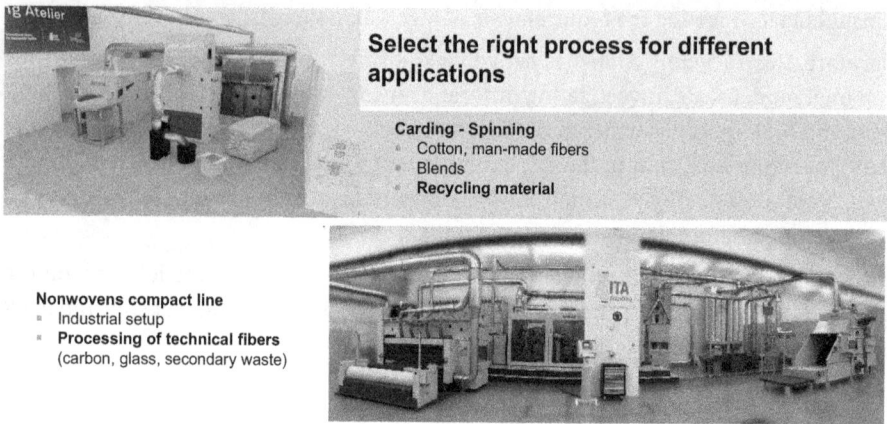

Figure 1.9 Process selection

The concept in Figure 1.10 shows the implementation of a closed material loop for technical textiles that focuses on the production of fiber-composite structures. This illustrates the effectiveness of the development methodology at conceiving a novel manufacturing strategy that uses waste generated from composites and has been successfully introduced into practice. The utilization of nonwoven fabric technology allows for an adapted and simultaneously cost-effective manufacturing concept for the finite-length fibers obtained from composites waste. Moreover, it facilitates the successful realization of products based on new applications in the field of fiber-composite materials through the material class that has been created: web-based composites.

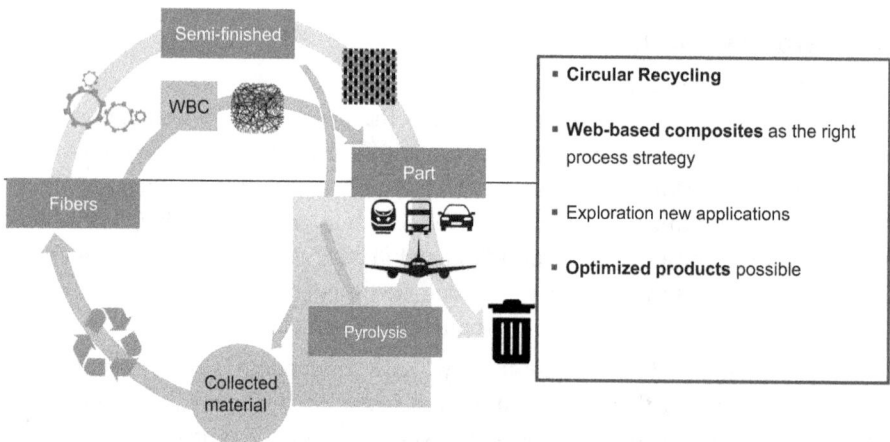

Figure 1.10 Cradle-to-cradle material flow with web-based composites (WBCs)

1.3.4.3 Design for Recycling

Not in every application does direct recycling lead to equivalent and high-quality products, and so increased attention needs to be paid to design for recycling. Despite a clear and thorough analysis of qualities and process capabilities, often no economically or technically viable means of direct reuse emerges. However, insights from the model factory can be used to derive requirements for adapting future products to better design for recyclable new products.

The reasons behind this are that recycled textile raw materials frequently have limited lengths, which tend to decrease with each processing step; this eventually yields a significantly broader distribution of fiber lengths. Furthermore, the various fibers used in textile products complicate recycling and, after appropriate processing such as mechanical recycling, lead to noticeably lower material quality, especially in terms of length. As a consequence of the significantly lower quality, either a redesign of new products (e.g., through altered material use) or a more suitably adapted processing approach is needed.

Using nonwoven fabric processing as an example, the following figure illustrates this concept, which leverages the high volume production of nonwoven fabric machines for improved cost-effectiveness. However, Figure 1.11 also demonstrates how choosing a different product can positively influence the quality of textiles produced through the nonwoven fabric process.

Examples	Standard fibers	Special fibers
Downcycling	Carpet backing, painting webs, pond webs, isolating material	Cleaning towels, door covers
Upcycling	Functionalized nonwovens (acustics, construction)	Seating systems, battery cases, aviation side panels

Figure 1.11 Upcycling using nonwoven production

1.3.4.4 Upscaling

The examples above have illustrated the application of a model factory for the recycling of textile materials, particularly in terms of expediting the development of suitable high-quality products. However, transferability to the scaling of industrial production must be appropriately considered with each new product development and be ensured through gradual scaling to production scale. For a rough product characterization, sample sizes of a few kilograms of material may be sufficient to provide an initial assessment of product development. In the next step, production sizes of approximately 100 kilograms can be used to derive initial insights for industrial realization. With the help of industrial partners integrated into the Recycling Atelier, transfer to an industrial production facility can then be ensured in a short time.

1.3.4.5 Learning Factory

Product development based on secondary raw materials from used textiles is considerably more complex than the production of standard textile items, which already relies on extensive experience. Additionally, knowledge of the development of suitable process steps has been partially lost in many industrialized countries. Therefore, the successful realization of a new development concept for recyclable textile products from used textiles also involves conveying the new insights in an appropriate form to various stakeholders in a future textile circular economy.

Target audiences for training and knowledge dissemination include professionals in textile production and future trainees in scientific and industrial education and training. Especially considering that users need to be more involved in the realization and application of new textiles from secondary raw materials, the focus here should be on targeted knowledge transfer to textile users. Given the significant changes in textile purchasing behavior as a result of online shopping and the widespread use of new electronic media, applied learning and training concepts must consider new learning methods and modern, intelligent knowledge transfer concepts. This can be achieved, for example, through the application of virtual reality and modern gaming concepts. Consequently, a learning factory for textile recycling was incorporated into the realization of the Recycling Atelier.

1.3.5 Summary

The implementation of a textile circular economy imposes greater demands on product development. The complexity is further increased by the use of secondary raw materials, and additional requirements arise from the circular economy. New tools are therefore needed for the development of textiles. Using the example of a model factory for textile recycling, this chapter has demonstrated with actual examples how such approaches may help existing deficits to be covered with new development tools, i.e.:

- In sorting, the use of automatic sorting based on the additional use of artificial intelligence will result in better sorting quality and reduce labor costs to allow application in industrial setups.

- Variable process selection makes it possible to find the best balance of technical, economic and ecological fulfillment of product properties.

- Design for recycling very often is mandatory to allow high-quality properties in the final product made from recycled raw materials.

- Upscaling to industrial applications needs to be realized in a stepwise manner to allow the transfer of findings in the model factory to high volume production.

- As recycling requires additional knowledge to be implemented in the design procedure and the process setup, the aspect of integrated training and education needs to be considered so as to ensure that new research findings can be implemented without a lack of proper understanding.

References for Section 1.3

[1] Bundesverband Sekundärrohstoffe und Entsorgung e.V. (bvse): Textilstudie 2020: Bedarf, Konsum, Wiederverwendung und Verwertung von Bekleidung und Textilien in Deutschland (2020), p. 8–11 & p. 23ff.

[2] Schlichter, S.: Modellwerkstatt Recycling – eine Antwort auf die Herausforderungen der textilen Kreislaufwirtschaft, 9. Internationaler Alttextiltag in Amsterdam 2022

[3] Ellen MacArthur Foundation, 2017: A new textiles economy: Redesigning fashion future

[4] European Commission; EU Strategy for sustainable and circular textiles, COM (2022), p. 141ff.

2
Political Background

2.1 Sustainability-Oriented European Legislation from the Perspective of Associations and Initiatives

Mauro Scalia, EURATEX, European Apparel and Textile Confederation, Felix Merkord, ITA Augsburg gGmbH

Europe faces a significant textile waste issue, with McKinsey's 2022 study revealing around 7 to 7.5 million metric tons of waste annually, highlighting a need for increased collection and improved recycling technologies to establish a circular textile economy and mitigate environmental impact. Closing the textile loop requires the establishment and scaling of five key steps [1]:

1. Collection
2. Sorting for reuse
3. Sorting for recycling
4. Preparation
5. Recycling

Europe could make significant progress by 2030 by creating 150–250 facilities in the areas of sorting for reuse and recycling, mechanical recycling, thermomechanical recycling, chemical recycling, and thermo-chemical recycling. This in turn could create up to 15,000 new jobs. According to the study, the most critical factors influencing the success of this project are [1]:

- Sufficient scaling
- Intensive cooperation [4]
- Adequate transition funding from industry

- Government and private sector investment
- High public pressure

2.1.1 Current EU Legislation

To achieve the desired goal of a closed circular textile economy in Europe, incentives need to be created for the industry. Among other things, this requires a comprehensive adaptation of national and international legislation. In the context of the Green Deal, which the European Union adopted in 2020 with the aim of becoming the first climate-neutral continent, the focus is not only on the zero-pollution target, but also on healthier food, the protection of biodiversity, sustainable mobility, environmentally friendly construction, and clean energy. A key objective here is the transformation of industries away from linear economic processes toward closed circular economy systems [5].

Regarding the textile and clothing industry, currently there are 16 legislative procedures in preparation that will have a significant impact (see Table 2.1). These legislative measures are part of the Green Deal efforts and are intended to create the necessary legal framework to encourage the industry to use resources and waste more sustainably. The planned legislative changes are expected to make a decisive contribution to promoting the circular economy in this sector [6].

Table 2.1 Current EU Legislation Concerning the Textile and Clothing Industries

Directive	Summary
Eco-Design for Sustainable Products Regulation	This proposal outlines a structure for defining eco-design requirements tailored to distinct product categories, aiming to notably enhance their circularity, energy efficiency, and other environmental sustainability dimensions. This initiative facilitates the establishment of performance and information standards applicable to nearly all types of tangible goods available in the EU market, with specific exclusions, like food and feed [7].
Extended Producer Responsibility (EPR)	Extended Producer Responsibility (EPR) is a policy instrument that broadens the financial and/or operational obligations of the producers of a product to encompass its handling in the post-consumer phase. This measure is implemented to contribute to the achievement of national or EU recycling and recovery targets [8].
Waste Shipments	The surge in economic growth and globalization has resulted in a global upswing in the cross-border transportation of waste, facilitated by road, rail, and maritime routes. In response, the EU has established a framework to oversee and regulate the movement of waste within its boundaries, as well as with nations that are parties to the Basel Convention. Additionally, the EU has implemented regulations governing the export, import, and intra-EU shipment of plastic waste [9].

Directive	Summary
Green Claims and Textile Labeling	As per this proposal, companies opting to assert "green claims" about their products or services will be obliged to adhere to specified minimum standards concerning the substantiation and communication of such claims. The proposal specifically addresses unequivocal assertions, such as "T-shirt made from recycled plastic bottles," "100% CO_2 compensated delivery," "Packaging comprises 30% recycled plastic," or " climate neutral sunscreen." It also seeks to address the proliferation of labels, including both new public and private environmental labels [10].
Green Public Procurement (GPP)	Green Public Procurement (GPP) is characterized as "a procedure in which public authorities aim to acquire goods, services, and works with a diminished environmental impact throughout their life cycle, in comparison to goods, services, and works with the same primary function that would be otherwise procured." Although GPP is a voluntary instrument, allowing member states to determine the extent to which policies or criteria are applied, it assumes a pivotal role in the EU's endeavors to enhance a resource-efficient economy [11].
Waste Legislation	The EU's waste policy seeks to advance the circular economy by maximizing the extraction of high-quality resources from waste. Aligned with the objectives of the European Green Deal, which strives to foster growth through the shift to a contemporary, resource-efficient, and competitive economy, there is a plan to reassess various EU waste laws as part of this overarching transition [12].
Corporate Sustainability Due Diligence	The primary objective of this Directive is to cultivate sustainable and responsible corporate conduct, embedding considerations for human rights and the environment into companies' operations and corporate governance. The proposed rules are designed to ensure that businesses systematically address the adverse impacts of their actions, both within and outside Europe, throughout their value chains [13].
Corporate Sustainability Reporting Directive	EU legislation mandates that large companies, as well as all listed companies (excluding listed micro-enterprises), divulge information regarding their perceptions of risks and opportunities associated with social and environmental matters. Additionally, they are required to report on the impact of their activities on both people and the environment. This disclosure serves as a valuable tool for investors, civil society organizations, consumers, and other stakeholders, and facilitates the assessment of a company's sustainability performance [14].

Table 2.1 Current EU Legislation Concerning the Textile and Clothing Industries (*continued*)

Directive	Summary
EU Taxonomy for Sustainable Activities	The EU taxonomy stands as a foundational element within the EU's sustainable finance framework and serves as a crucial tool for market transparency. Its primary function is to guide investments toward economic activities essential for the transition, aligning with the objectives of the European Green Deal. Functioning as a classification system, the taxonomy establishes criteria for economic activities that are in harmony with a net-zero trajectory by 2050, encompassing broader environmental goals beyond climate considerations [15].
PFAS Restriction Proposal	This proposal seeks to reduce emissions of PFAS to the environment and enhance the safety of products and processes for individuals [16].
Skin Sensitizers	The European Chemicals Agency (ECHA) and member states distinguish skin sensitizers by analysing registration data provided by the industry under the REACH Regulation (Registration, Evaluation, Authorization, and restriction of Chemical substances). This data, combined with information from various other sources, undergoes thorough screening. Upon identification of a new potential substance of concern, appropriate regulatory risk management measures will be implemented [17].
Bisphenol A	In April 2023, the European Food Safety Authority (EFSA) released a revised safety assessment of BPA in Food Contact Materials (FCM), notably reducing the tolerable daily intake (TDI) level established in its previous evaluation conducted in 2015 [18].
REACH Revision	The regulation concerning the registration, evaluation, authorization, and restriction of chemicals (REACH) is the primary EU legislation designed to safeguard human health and the environment against potential risks posed by chemicals. This is achieved through improved and early identification of the inherent properties of chemical substances and the implementation of measures, including the phased reduction or restriction of substances deemed to be of very high concern [19].
PFHxA Restriction	A proposal for restriction can be formulated either by a member state or by the ECHA, either upon the request of the Commission or independently, particularly for substances listed in the Authorization List. It is a mandatory legal procedure for a member state to inform the ECHA of its intent to compile a restriction dossier [20].

2.1.2 Influence of Associations and Initiatives

The development of key legislative initiatives is heavily supported by the involvement of external organizations, associations, and initiatives. In the field of textiles and clothing, the central association looking after the industry's demands is EURATEX, the European Apparel and Textile Confederation. This is the representative body for the European textile and clothing industry within the EU institutions. Functioning as the voice of the European industry, it strives to establish an environment within the European Union conducive to facilitating the manufacturing of textile and clothing products. Associations such as EURATEX act as key players and provide advice in the interests of European small and medium-sized enterprises (SMEs) in particular. EURATEX works proactively with EU institutions and policy makers. In addition, EURATEX is involved in various initiatives aimed at tackling the growing challenge of attracting and retaining qualified human resources for the EU textile and clothing industry [21].

2.1.2.1 Examples of Regulations Supported by EURATEX

Three examples of currently envisaged laws which have come about thanks in part to the work of EURATEX are the Eco-Design for Sustainable Products Regulation (ESPR), the Digital Product Passport (DPP) and the Green Claims Directive [6]. The laws are described in the following.

2.1.2.1.1 Eco-Design for Sustainable Products Regulation (ESPR)

The Eco-Design for Sustainable Products Regulation (ESPR) provides for the introduction of new minimum requirements for the sale of products and defines mechanisms for conformity assessments and verifications. Various EU stakeholders are involved in this process, including the Commission, the Parliament, and the Council, as well as non-governmental organizations and industry. EURATEX has put forward preliminary proposals aimed at creating a level playing field and supporting small and medium-sized enterprises (SMEs), with a focus on the clothing sector. The proposed minimum requirements focus on three fundamental aspects: durability and quality, reuse, and recycled content. The legislative proposal passed the EU institutions in 2023 and the final version is currently being prepared for publication in 2024 [6].

2.1.2.1.2 Digital Product Passport (DPP)

The digital product passport (DPP) provides for mandatory data sharing for consumers and public authorities and involves various stakeholders at EU level, including the Commission, the Parliament, non-governmental organizations, and industry. EURATEX has put forward preliminary proposals that target aspects such as interoperability, information selection, traceability, and implementation costs, particularly regarding small and medium-sized enterprises (SMEs). The legislative proposal was endorsed by the EU institutions in 2023 and the final version was being prepared for publication in 2024 [6].

2.1.2.1.3 Green Claims Directive

In future, companies will have to justify sustainability claims using a comprehensive methodology that includes references to methods, such as the product environmental footprint (PEF) and new labels. In its preliminary proposals, EURATEX has emphasized its willingness to support Europe-wide harmonization through PEF while calling for improvements to the PEF system. The legislative proposal was adopted by the EU institutions in 2023 and the final version is currently being prepared for publication in 2024 [6].

2.1.2.1.4 Product Environmental Footprint Category Rules (PEFCR)

EURATEX has made a significant contribution to the development of the product environmental footprint (PEF) concept, whose product environmental footprint category rules (PEFCRs) were adopted by the EU for footwear and apparel textiles in 2024. The PEF methodology is based on the life cycle assessment (LCA) approach, a scientifically validated method for measuring the environmental impact at each stage of a product's life. Aside from 26 designated representatives of the textile industry, the European Commission and the European Environmental Federation (EEB) are also involved as observers. The data to be provided by companies for the PEFCR is divided into two main categories:

- Primary data: This includes company-specific information that comes either directly from the companies themselves or indirectly from suppliers. This data is specific to the product and supply chain and therefore forms a key basis for the assessment.

- Secondary data: Industry-standard averages, such as statistics or other published production data which are not collected, measured or estimated directly by the company concerned. The PEFCR can provide default values if required, and general data sets can be obtained from the EU Commission's EF 3 database for environmental footprints. This secondary data serves as additional reference points and enables a comprehensive assessment of environmental impact.

The chronological development of the PEFCR for apparel and footwear spans different phases. Starting in 2022, supporting studies were performed, and a testing phase of the PEFCR on real products by companies to evaluate its applicability and effectiveness was successfully completed. In 2023, a second public consultation was held in which stakeholders had the opportunity to express their opinions and suggestions. In 2024, the Technical Advisory Board (TAB) and the EU Subgroup (EU member states) made recommendations to the European Commission on the PEFCR for clothing and footwear. During this phase, standards were examined and defined, and possible legal applications were agreed. Later, the European Commission finally adopted the PEFCR for clothing and footwear. These regulations will serve as a reference for future legislation, where necessary. Brands and industry are encouraged to use this common framework to fully assess the environmental footprint of their products.

2.1.3 Impact of Initiatives

An example of the influence of associations such as EURATEX is the ReHubs initiative, a measure launched by the European Apparel and Textile Federation (EURATEX) in 2020 to tackle the immense problem of textile waste in Europe. The ReHubs initiative consists of three main groups of stakeholders [22]:

- Business Council: Pioneering companies have come together to carry out the TES (Textile Waste Collection and Sorting) study.

- Stakeholder Forum: This forum comprises a wider circle of players from business, research, and science. It has already met twice to exchange information at the highest level and to support future cooperation.

- Euratex Task Force: Comprising 14 national associations, this task force reviews the progress of ReHubs initiatives and coordinates with policy and industry developments at national level.

The initiative was launched to promote cooperation throughout the entire extended textile value chain. All perspectives in the areas of chemicals, fiber and textile manufacturing, garment production, retail, and distribution as well as the collection, sorting and recycling of textile waste are considered. ReHubs initiative aims to enable fiber-to-fiber recycling of 2.5 million tons textile post-consumer waste in European Union by 2030. The first project supported by ReHubs, "Transform Waste into Feedstock," plans to build a plant with a capacity of 50,000 tons by 2024, led by Texaid AG. This will focus on the further development and scaling of sorting technologies. In addition to this project, three further aims were announced [22]:

- Increase the acceptance of mechanically recycled fibers in value chains

- Expand capacity by solving technical challenges in thermomechanical textile recycling

- Create a capsule collection with recycled end-of-life products (valorization of waste from the Murano furnaces)

2.1.4 Summary and Conclusion

Europe faces a significant textile waste issue, with McKinsey's 2022 study revealing around 7 to 7.5 million tons of waste annually. This underscores the urgent need to enhance collection and recycling methods for a circular textile economy, as current practices fall short of effectively managing this waste stream. The establishment of five key steps is crucial to closing the textile loop: collection, sorting for reuse, sorting for recycling, preparation and recycling. These steps require concerted efforts and investments to scale up infrastructure and technology, aiming to capture more textile waste and channel it back into the production cycle. Moreover, the potential benefits of such a transition are substantial. By 2030, scaling up sorting and recycling facilities could create up to 15,000 new jobs, providing economic incentives for industry participation. Legislative measures, part of the EU's ambitious Green Deal agenda, play a pivotal role in driving this transformation. Initiatives like the Eco-Design for Sustainable Products Regulation and the digital product passport aim to set standards and facilitate transparency in the industry's environmental footprint, encouraging sustainable practices and innovation.

In this landscape, associations such as EURATEX play a crucial role in advocating for industry-wide cooperation and supporting initiatives to tackle textile waste challenges. EURATEX's ReHubs initiative, launched in 2020, exemplifies this collaborative approach, bringing together stakeholders across the textile value chain to promote innovation and sustainability. Through projects such as "Transform Waste into Feedstock," ReHubs aims to develop and scale up sorting technologies and thus to pave the way for increased acceptance of recycled fibers in value chains and address technical challenges in textile recycling.

Overall, the transition to a circular textile economy requires not only legislative support but also industry collaboration and technological innovation. By leveraging these resources and working toward common goals, Europe can mitigate the environmental impact of textile waste while unlocking economic opportunities and fostering a more sustainable future for the industry.

References for Section 2.1

[1]   Hedrich, S., Janmark, J., Langguth, N., Magnus, K.-H., Strand, M., *Scaling textile recycling in Europe-turning waste into value*: Textile recycling can turn Europe's textile waste into value and build a sustainable and profitable new industry. (2022)

[2]   Preuss, S., *Euratex' ReHubs-Initiative will massives Textilabfallproblem in Europa angehen* (2022), *www.fashionunited.de* (cited 02/24)

[3]   Preuss, S., *Studie zum Textilrecycling: „Europa wird den Weg weisen"* (2022), *www.fashionunited.de* (cited 02/24)

[4]   van Duijn, H., Carrone, N.P., Bakowska, O., Huang, Q., *Sorting For Circularity Europe: An Evaluation And Commercial Assessment Of Textile Waste Across Europe* (2022), *www.reports.fashionforgood.com*

[5]   European Union, *The European Green Deal*: Striving to be the first climate-neutral continent (2019), *www.commission.europa.eu* (cited 02/24)

[6]   M. Scalia, *EU Legislations and EURATEX* (2023)

[7]   European Union, *Eco-Design for Sustainable Products Regulation*: The new regulation will improve EU products' circularity, energy performance and other environmental sustainability aspects. (2022), *www.commission.europa.eu* (cited 02/24)

[8]   European Union, *Extended Producer Responsibility* (2023), *www.era-comm.eu* (cited 02/24)

[9]   European Union, *Waste shipments*: EU rules on transporting waste within and beyond EU borders, to protect the environment and public health. (2023), *www.environment.ec.europa.eu* (cited 02/24)

[10] European Union, *Consumer protection*: enabling sustainable choices and ending greenwashing (2023), *www.ec.europa.eu* (cited 02/24)

[11] European Union, *Green Public Procurement*: Procuring goods, services and works with a reduced environmental impact throughout their life cycle. (2023), *www.green-business.ec.europa.eu* (cited 02/24)

[12] European Union, *Waste and recycling*: Managing waste in an environmentally sound manner and making use of the secondary materials they contain are key elements of the EU's environmental policy. (2023), *www.environment.ec.europa.eu* (cited 02/24)

[13] European Union, *Corporate sustainability due diligence*: Fostering sustainability in corporate governance and management systems. (2022), *www.commission.europa.eu* (cited 02/24)

[14] European Union, *Corporate sustainability reporting*: EU rules require large companies and listed companies to publish regular reports on the social and environmental risks they face, and on how their activities impact people and the environment. (2023), *www.finance.ec.europa.eu* (cited 02/24)

[15] European Union, *EU taxonomy for sustainable activities*: What the EU is doing to create an EU-wide classification system for sustainable activities. (2023), *www.finance.ec.europa.eu* (cited 02/24)

[16] European Chemicals Agency, *ECHA publishes PFAS restriction proposal* (2023), *www.echa.europa.eu* (cited 02/24)

[17] European Chemicals Agency, *Skin sensitising chemicals* (2022), *www.echa.europa.eu* (cited 02/24)

[18] European Food Safety Authority, *Bisphenol A* (2023), *www.efsa.europa.eu* (cited 02/24)

[19] European Union, *REACH Regulation*: To protect human health and the environment against the harmful effects of chemical substances. (2023), *www.environment.ec.europa.eu* (cited 02/24)

[20] European Union, *Registry of restriction intentions until outcome* (2020), *www.echa.europa.eu* (cited 02/24)

[21] EURATEX, *About EURATEX* (2022), *www.euratex.eu* (cited 02/24)

[22] ReHubs, *Circulating Textile Waste into Value* (2023), *www.rehubs.eu* (cited 01/24)

2.2 EU Strategy for Sustainable and Circular Textiles: Legal Requirements & Consequences for Sustainability of Textiles

Fatah Naji, Wolfgang Rommel, bifa Umweltinstitut, Augsburg

2.2.1 The Vision until 2030

2019 saw the publication of the European Green Deal, a European climate pact that aims to transform the EU economy in a sustainable way so that climate protection targets can be achieved by 2030 and 2050.

Textiles are the fourth-largest consumers of primary raw materials and water (after food production, housing and transport) and the fifth-largest emitters of greenhouse gases [3]. In the EU Circular Economy Action Plan from 2020, used textiles were therefore defined as part of the central product value chains. A textile strategy was developed as a key measure under the title "Strategy for sustainable and circular textiles" and was published on March 30, 2022 [4].

The aim of the strategy is to transform the EU textile market so that consumers benefit from high-quality and affordable textiles, fast fashion goes out of fashion and profitable reuse and repair services become more widespread.

Furthermore, producer responsibility is to be strengthened for textiles along the entire value chain from raw material extraction to waste generation. In order to create the conditions for a functioning circular system, sufficient capacity for fiber-to-fiber recycling must therefore be built up and the disposal of textile waste by incineration and landfilling must be reduced to a minimum. As a result, textile products placed on the EU market will have to be more durable, recyclable, made largely from recycled fibers, free of hazardous substances, and social equity and justice as well as environmental protection must be in compliance with the law [5].

To ensure this, laws, regulations and directives have been passed or are in planning. At national level, individual member states already have regulations for a sustainable, circular textile economy.

The aim is to close the textile cycle, taking into account all the links in the chain. This means that the legal requirements for a sustainable textile industry must be developed for all stakeholders, such as manufacturers, retailers, consumers, waste collection, waste treatment and recycling (Figure 2.1). In the context of Extended Producer Responsibility (EPR), stakeholders take on different roles in the circular economy.

Figure 2.1 The textile cycle

2.2.2 Eco-Design and Digital Product Passport

Eco-Design Directive 2009/125/EC "Establishing a framework for the setting of eco-design requirements for energy-related products"creates a framework for the setting of eco-design requirements and regulations for certain product groups placed on the EU market. Around 40 product groups are currently affected, such as washing machines, refrigerators and TVs. Not only energy-consuming products are affected, but also products that influence the energy consumption of other systems, such as double-glazed windows. The aim is to increase the recyclability, energy efficiency and ecological sustainability of these products. One exception is food and animal feed, for example, which are defined in Regulation 178/2002.

The Eco-Design Directive is not a waste regulation, but has an impact on the entire life cycle of products. It does not contain any specific regulations, but authorizes implementing measures, such as EU regulations, and includes the possibility of self-regulatory initiatives by the industry because each product category requires different preconditions.

The European Commission's draft Sustainable Products Initiative (SPI) and the EU strategy for sustainable textiles propose extending the Eco-Design Directive to other product groups. These include steel, cement, chemicals, furniture and textiles. It is therefore to be expected that the requirements for textiles in terms of sustainability criteria, durability and recyclability will be more stringent in the future. For example, a higher content of recycled fibers is currently being discussed.

Further requirements will affect the quality of textile products. They relate to a longer product life, reusability, upgradability and repairability of textile products. However, the criteria and testing methodology for this have not yet been developed. Substances that have a negative impact on the recyclability of textiles during production are also to be excluded. Even if the requirements here have not yet been defined, this could lead to the exclusion of certain coatings. The carbon footprint and ecological footprint of textile products will also play an important role.

2.2.3 Labeling and Information Requirements for Textiles

In the future, requirements will be imposed on product information and this will lead to the further development of the labeling and information obligation and ultimately to the introduction of a digital product passport for textile products that will allow stakeholders, such as consumers, to make ecologically-based purchasing decisions and to act responsibly.

With regard to recycling-friendly design, the EU Textile Labeling Regulation (EU Regulation 1007/2011) regulates the designations of textile fibers and the associated labeling and marking of the fiber composition of textile products and thus also the labeling of textiles. Textile products and textile fibers are defined in Article 3 as follows:

- A textile product is therefore "a product that contains only textile fibers in the raw, semi-processed, processed, semi-manufactured, processed, semi-manufactured or manufactured state, regardless of the process used to mix or combine them;"

- A "textile fiber" is "a product characterized by its flexibility, its fineness and its great length in relation to its maximum cross-sectional area and thus suitable for the manufacture of textile products, or a flexible strip or tube of a normal width not exceeding 5 mm, including strips cut from wider strips or sheets, manufactured from the materials used for the manufacture of the fibers listed in Table 2 of Annex I and suitable for the manufacture of textile products."

The fibers listed in Table 2 of Annex I include viscose, silk, cupro, polyamide and elastane, as well as glass fibers. The textile labeling requirement applies to the following products:

- "Products containing at least 80% textile fibers by weight

- Covering material for furniture, umbrellas and parasols with a textile component content of at least 80% by weight

- The textile components of the top layer of multilayer floor coverings, of mattress covers and of covers of camping articles, provided that these textile components account for at least 80% by weight of these top layers or covers

- Textiles that are incorporated into other goods and become part of them, provided their composition is indicated"

However, the Eco-Design Directive and the digital product passport (DPP) introduced in this context are intended to provide even more information about the entire life cycle of textile products (Figure 2.2). With the DPP, the European Commission is pursuing the goal of promoting sustainable products and enabling the transformation of the linear economy into a circular economy. Consumers are also to be given the opportunity to take responsibility by empowering them to make buying decisions that are sustainable, which is only possible with sufficient information. In addition, the digital product passport is intended to create transparency so that authorities will be able to check whether companies are complying with legal obligations for their products.

The implementation of the DPP in textile products could be realized through the use of interfaces such as radio frequency identification (RFID), QR codes and near-field communication (NFC). Which data and information must be stored and which system is most suitable is currently being investigated and evaluated at European level.

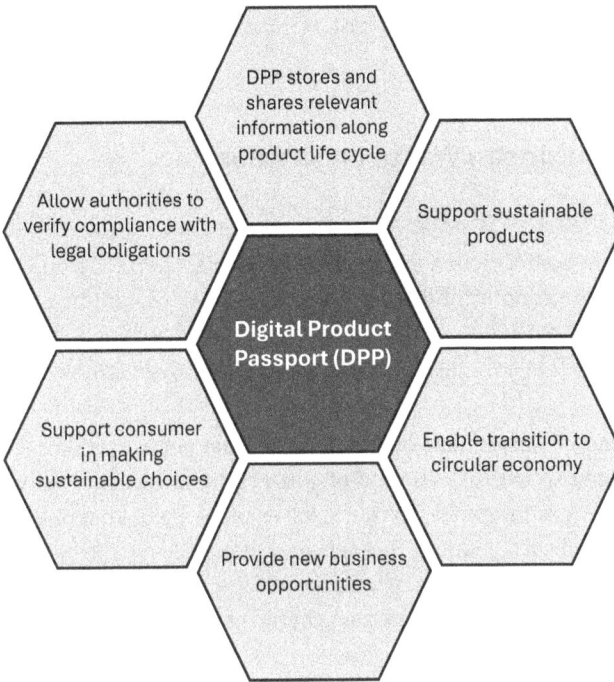

Figure 2.2 Objectives of the DPP concept; adapted from [7]

2.2.4 Duty of Care for Textile Products

Another EU objective is to ban the destruction of unsold and returned textile goods. This is intended to minimize textile waste on the one hand and promote responsible storage concepts for textile manufacturers and retailers on the other. Corresponding laws already exist in individual member states.

For example, France has enacted the "anti-gaspillage et économie circulaire (AGEC)," a law against waste and for a more circular economy. This law now prohibits the destruction of unsold products.

In Germany, the duty of care is enshrined in the Circular Economy Act, but is currently only a non-binding or passive obligation. An enforceable obligation only exists if the duty of care is specified by ordinance and can therefore be implemented (ordinance authorizations) [1]. Recovery from textile waste is only permitted if the original purpose of the textiles can no longer be maintained, other uses are not possible, reuse is no longer possible or is no longer economically reasonable. In addition, manufacturers and retailers may be required to submit a transparency report detailing the type, quantity, whereabouts and disposal of textiles as well as the measures taken and planned to implement the duty of care (LAGA M40 2023).

The question therefore arises as to which characteristics must be fulfilled for textiles to reach the end-of-life status.

2.2.5 Textile Waste in European Waste Legislation

2.2.5.1 Textile Products and End of Life

Waste legislation only comes into effect when textiles become waste. Therefore, one of the key questions in the circular economy of textiles is the question of the product status and the transition to waste. Legislation in Germany has already developed definitions for this, and can be expected in a similar form in the amended Waste Framework Directive of the EU.

A possible definition of the end-of-waste status of textile waste is set out in Section 5 (1) of the German circular economy act (KrWG). The end-of-waste status is deemed to have been reached when textile waste is prepared for reuse, for example by having undergone a recycling process and is of such a nature that certain criteria are met. Recycling can be achieved, for example, through sorting, testing/inspection, cleaning or repair. The used textiles must then meet the following product characteristics:

- "Normally used for specific purposes

- There is a market for them or a demand for them

- They meet all the technical requirements applicable to their intended purpose and all legal provisions and applicable standards for textiles, and

- Their overall use does not lead to harmful effects on humans and the environment"

According to German legislation, the contents of used clothing collection containers are by definition waste. Only used textiles that have been sorted, checked/inspected, cleaned or repaired before being made available for reuse are considered second-hand goods. The resulting textile residues from mechanical waste treatment are waste with waste code number 19 12 08 until new products are created from them, such as cleaning cloths.

In some EU member states, after preparation for reuse, the ready-to-wear and marketable textiles may contain other substances or even textiles that still fulfill waste characteristics by definition. This includes, for example, foreign substances or harmful substances. Otherwise, the entire batch or bale remains subject to waste legislation. The question of waste status and product status is particularly important with regard to the import and export of textiles made from waste.

2.2.5.2 EU Waste Framework Directive

On July 5, 2023, the EU Commission published its proposal for a partial revision of the Waste Framework Directive (2008/98/EC). It proposes introducing a binding and harmonized extended producer responsibility for textiles. To this end, requirements for product design, for example, are to be developed. The extended producer responsibility is to be introduced and measures to combat greenwashing are to be introduced. In addition, guidelines to reduce overproduction and overconsumption are to be implemented. Furthermore, the destruction of unsold or returned textiles, which is currently still common practice in many EU member states, is to be stopped. The protection of the oceans from emissions of microplastics from synthetic textiles is also to be given prominence. Sustainability is to be strengthened by creating incentives for circular business models, for example through reuse and repair. The export of textile waste outside the EU is also to be restricted.

Article 11 of the Waste Framework Directive stipulates that, in addition to paper, metal, plastic and glass, EU member states must also collect textiles separately from January 1, 2025.

Under Article 6 of the Waste Framework Directive, each member state must take measures to amend the provisions on the end-of-waste status of textile waste. This includes developing criteria to define when textiles cease to be waste.

Article 3 of the same directive defines municipal waste as "mixed waste and separately collected waste from households, including paper and cardboard, glass, metal, plastic, biowaste, wood, textiles, packaging, waste electrical and electronic equipment, spent batteries and accumulators, and bulky waste, including mattresses and furniture."

The European waste catalog was defined in the EU for the purpose of precision and harmonization of waste designations.

2.2.5.3 European Waste Catalog – Waste List

The European waste catalog (EU Regulation 2000/532/EC) categorizes textile waste according to the following chapters and groups. Post-production textile waste includes waste from composite materials, such as impregnated textiles, elastomers, plastomers (04 02 09), waste from untreated textile fibers (04 02 21) and waste from processed textile fibers (04 02 22). Textiles from the mechanical treatment of waste are summarized under waste code 19 12 08. Post-consumer textile wastes are also listed. These include clothing (20 01 10) and other textiles (20 01 11), which are counted as municipal waste under Chapter 20. Further waste codes for textile waste are listed in Table 2.2.

Table 2.2 Textile Wastes and their Waste Codes According to the European Waste Catalog

Waste code	
04 02 09	Waste from composite materials (impregnated textiles, elastomer, plastomer)
04 02 10	Organic substances from natural materials (e.g. fats, waxes)
04 02 14*	Waste from finish-containing organic solvents
04 02 15	Wastes from finishing operations other than those mentioned in 04 02 14

Table 2.2 Textile Wastes and their Waste Codes According to the European Waste Catalog (*continued*)

Waste code	
04 02 16*	Dyes and pigments containing hazardous substances
04 02 17	Dyes and pigments other than those mentioned in 04 02 16
04 02 19*	Sludge from in-house wastewater treatment containing hazardous substances
04 02 19	Sludges from on-site waste water treatment other than those mentioned in 04 02 19
04 02 21	Waste from untreated textile fibers
04 02 22	Waste from processed textile fibers
04 02 99	Waste not otherwise specified
15 01 09	Packaging made from textiles
19 12 08	Waste from the mechanical treatment of waste – here textiles
20 01 10	Municipal waste, separately collected fractions – here clothing
20 01 11	Municipal waste, separately collected fractions – here textiles

2.2.5.4 Waste Shipment Regulation

In the case of transboundary waste shipments, the question always arises as to whether waste can be exported without the involvement of authorities or whether a notification procedure is required. This depends on the classification of the waste under different multi-national conventions. In addition to the Waste Shipment Regulation, the Basel Convention and the OECD Council Decision also apply to EU member states.

If a waste is listed as a green waste, no involvement of authorities is generally required, apart from economic authorities (e.g. customs). If a waste is listed as amber waste or is not listed, a notification procedure must be carried out.

For homogeneous, used textiles destined for recycling (without contamination), Basel code B3030 of the Green Waste list applies. If textile waste on the Green Waste list or mixtures of used textiles as defined in Annex IIIA of the WSR are contaminated with other materials such that they exhibit hazardous properties or environmentally friendly recovery is prevented, this waste must be assigned to the Amber Waste list.

This applies, for example, to waste contaminated with organic or inorganic harmful adhesions, such as waste contaminated with oil, solvents or heavy metals, cleaning cloths, absorbent cloths or protective clothing. If no Basel or OECD code applies to certain textile waste, it is considered unlisted waste that must undergo a notification procedure prior to export.

In the future, stricter regulations for the export of waste to non-OECD countries are to be implemented for transboundary waste shipments in accordance with Vision 2030. In addition, the monitoring and enforcement of regulations for waste shipments to OECD countries are to be improved. When exporting waste, companies in the EU that export waste from the EU are to be obliged to ensure that the facilities in the recipient country are subjected to an independent audit. These regulations will also affect exports of textile waste to sorting plants and mechanical treatment plants in order to ensure that textile waste is also recycled in an environmentally friendly manner in non-European countries and is not disposed of illegally. In this context, sanctions for illegal shipments are also to be tightened in future.

Stricter rules are to apply to the shipment of waste for landfill or incineration. The aim is to only approve these in a few, well-founded cases.

At the same time, the administrative procedures for the recovery of waste within the EU are to be simplified in order to strengthen the circular economy in the EU. This also includes fully digitizing the notification procedure under the WSR.

However, the EU also wants to become active at an international level. For example, by promoting measures at international level to improve waste management and sustainability in the global waste trade.

In order to combat illegal waste shipments, the European Commission wants to take stronger measures against the illegal waste trade. To this end, the European Commission intends to intensify cooperation with the European Anti-Fraud Office (OLAF) with regard to cross-border crime in relation to the illegal waste trade.

2.2.6 Collection of Textiles in the Context of Extended Producer Responsibility (EPR) for Textiles

Under the EU Waste Framework Directive, the separate municipal collection of used textiles must be introduced by the member states in the EU from 2025. Collection can take place in a wide variety of ways.

When collecting used textiles, a distinction must be made between pick-up and drop-off systems. In the case of pick-up systems, the collection containers are placed at the waste producer's premises, for example on private property, while in the case of drop-off systems, either municipal recycling centers or depot containers positioned on public streets are used.

The installation of depot containers in the public street space requires an urban planning concept and can be carried out by different players who may be in competition with each other:

- Collection under municipal responsibility
- Commercial collection

- Charitable collection

- Take-back by manufacturer or retailer

Depending on the player, collection can take place, for example, via drop-off systems at container locations, at recycling centers or by a company commissioned by the municipality or by collection directly from households.

Classic depot containers in public street areas are already widely used in the EU for glass and increasingly also for waste electrical and electronic equipment (WEEE). For the collection of used textiles, depot containers are considered one of the most suitable collection systems from both a qualitative and quantitative point of view and are therefore frequently used by public waste management authorities to fulfill their currently public textile disposal obligations [6].

The marketing of viable used textiles can generate revenue, which leads to collection containers or other collection bins being set up illegally. Therefore, a legal basis and control mechanisms need to be created to curb the illegal collection of used textiles and enable the proper collection of used textiles.

When developing collection concepts, it is always important to consider how many containers are needed and where the ideal locations are in order to achieve sufficiently efficient distribution. The emptying rhythm for container collection should also be based on the expected collection volume.

Experience has shown that sites for waste collection containers, not only for used textiles, quickly become illegal waste dumping and littering areas. This leads to neglect, which not only results in an unsightly cityscape, but also causes hygiene problems, for example when residual and organic waste and similar items are deposited. Other difficulties with textile collection using containers include

- Theft of high-value used textiles from the containers

- Incorrect waste that leads to the destruction of used textiles (e.g. used oil, paint, food)

- Moisture and dampness that leads to mold growth

- Fire hazard due to disposal of electrical appliances or batteries

- Arson by fireworks, cigarettes and the like

In the EU, extended producer responsibility is addressed to stakeholders who develop, manufacture, process, treat, sell, or import products on a professional basis. It therefore also applies to the textile industry. The aim of extended producer responsibility is to promote the design and manufacture of textiles throughout their entire life cycle without hindering the free movement of goods in the internal market. Textiles should have the following properties:

- Repairability

- Reusability

- Dismantlability

- Recyclability

- Resource efficiency

The member states must take measures to ensure that stakeholders assume financial responsibility after the textiles have been used, i.e. when taking back returned textiles and managing textile waste. In this way, requirements are established for the handling of waste and requirements for cost responsibility for waste disposal in accordance with the polluter pays principle. This also includes the obligation to provide publicly accessible information on reusability and recyclability.

Member states are also to take measures to minimize the environmental impact of waste generation. For example, this can be done by promoting textiles that are reusable, technically durable and can be disposed of in an environmentally friendly way once they reach the end of their life. However, the following criteria should be taken into account in all measures taken by member states:

- Technical and economic feasibility

- Overall impact on the environment and human health

- Social consequences, and

- The proper functioning of the EU internal market

The general minimum requirements for extended producer responsibility are defined in the Waste Framework Directive as follows [2]:

- Precise definitions as well as roles and responsibilities of all stakeholders involved

- Measurable targets in line with the waste hierarchy

- Reporting on quantities placed on the market and volume flow records

- Equal treatment of producers

- Information for waste owners affected by EPR systems and incentives to use the infrastructure introduced as part of the EPR system

In Germany, the Packaging Act requires operators ("systems") of glass containers or their contractual partners for the operational provision of services to set up a nationwide household-based return system for packaging (Section 14 (1) VerpackG). In return, they receive fees from manufacturers and distributors of packaging. As textiles are not packaging, this does not apply to them (Bundestag 05.07.2017).

Nevertheless, it is conceivable that this concept could also be transferred to textiles at EU level as part of extended producer responsibility. However, experience has shown that this would involve a great deal of effort.

In a study commissioned by the German Federal Environment Agency in March 2023, four possible models of extended producer responsibility or producer responsibility were developed and evaluated. In this study, a distinction was made between the "fund concept," "manufacturer-supported concept," "systems in competition concept" and "contract concept" [8]. Whether and, if so, which of the models will be implemented in Germany or in the EU is not certain at this time:

- Under the fund concept, manufacturers pay into a fund for the textiles they place on the market. The fees are staggered according to ecological criteria. The fund's resources can then be used to finance waste collection and disposal or recycling.

- Under the producer-driven concept, the organizational and financial responsibility for fulfilling the collection and recycling of textile waste lies in the hands of the manufacturers. Manufacturers can, for example, introduce a take-back scheme for used textiles at their own expense or set up a joint system for manufacturers. To this end, legal regulations must be defined in which the requirements for implementation are regulated and control instruments for authorities are developed

- The third concept provides for several collection systems in competition. These collection systems must be checked and approved by a competent authority. Manufacturers must then pay the collection systems for the collection and disposal of all their quantities placed on the market. At the same time, it is no longer possible for manufacturers to take back used textiles of their own brand, as the collection systems are responsible and in charge. However, there is a choice between several system operators, which makes for competition.

- Under the contract-based concept, manufacturers are obliged to conclude contracts with certified collectors, sorters and/or recyclers. These are then responsible for ensuring that the collection, sorting and recycling of a contractually agreed quantity of used textiles is carried out. All statutory requirements for collection, sorting and recycling must be met by the contractual partner. Organizational structures are not specified in this concept.

2.2.7 Summary

The European Green Deal is a comprehensive plan launched in 2019 to make the EU economy more sustainable and to reduce greenhouse gas emissions in two steps by 2030 and 2050. Textiles play a significant role in these efforts, as they are major consumers of raw materials and contributors to greenhouse gas emissions. To promote sustainability in the textile industry, the EU Strategy for Sustainable and Circular Textiles aims to improve the durability, recyclability, and environmental performance of textiles. This involves measures such as reducing textile waste, promoting producer responsibility and encouraging fiber-to-fiber recycling.

To support these goals, the Eco-Design Directive will be extended to include textiles, with a focus on sustainability, durability, and recyclability standards. The introduction of a digital product passport for textiles will provide consumers with detailed information about the product's life cycle, enabling them to make more sustainable choices.

EU regulations will also address the destruction of unsold textiles, thereby promoting responsible storage and waste minimization. The Waste Framework Directive will expand the responsibility of textile producers and stress sustainable design, reuse, and recycling. By 2025, EU member states will be required to collect textile waste separately and take steps to reduce illegal exports and improve textile recycling.

Extended producer responsibility schemes will involve stakeholders in financing waste management and promoting environmentally friendly product life cycles. The EU is exploring different models to implement these measures effectively, with a focus on minimizing environmental impacts and supporting the circular economy.

References for Section 2.2

[1] Bundestag (24.02.2012): Act to Promote the Circular Economy and Ensure the Environmentally Sound Management of Waste. Kreislaufwirtschaftsgesetz – KrWG, last amended by Art. 5 G v. 02.03.2023 I No. 56, *https://www.gesetze-im-internet.de/krwg/KrWG.pdf*, last checked on 01.11.2023.

[2] EUKOM (19.11.2008): Directive 2008/98/EC of the European Parliament and of the Council of November 19, 2008 on waste and repealing certain Directives. Waste Framework Directive 2008, *https://eur-lex.europa.eu/legal-content/DE/TXT/PDF/?uri=CELEX:02008L0098-20180705*, last checked on 01.01.2023.

[3] EUKOM (2020): A new Circular Economy Action Plan. A new Circular Economy Action Plan For a cleaner and more competitive Europe. Ed. by European Commission (EUKOM). EU-KOM, *https://eur-lex.europa.eu/resource.html?uri=cellar:9903b325-6388-11ea-b735-01aa75ed71a1.0016.02/DOC_1&format=PDF*, last checked on 01.11.2023.

[4] EUKOM (2022): EU Strategy for Sustainable and Circular Textiles. EU Strategy for Sustainable and Circular Textiles. Ed. by the European Commission (EUKOM). EUKOM, *https://eur-lex.europa.eu/resource.html?uri=cellar:9d2e47d1-b0f3-11ec-83e1-01aa75ed71a1.0013.02/DOC_1&format=PDF*, last checked on 01.11.2023.

[5] EUKOM (2023): EU strategy for sustainable and circular textiles. To create a greener, more competitive textiles sector. Ed. by European Commission (EUKOM), *https://environment.ec.europa.eu/strategy/textiles-strategy_en*, last checked on 01.11.2023.

[6] LAGA (2023): Communication Federal/Länder Working Group on Waste (LAGA) 40. Enforcement aid for the prevention and collection, sorting and recycling of used textiles. Estab lished by LAGA.

[7] Saari et. al (2022): Saari, L., Heilala, J., Heikkilä, T., Kääriäinen, J., Pulkkinen, A., & Rantala, T. (2022). Digital product passport promotes sustainable manufacturing: whitepaper. VTT Technical Research Centre of Finland.

[8] UBA (2023): Development of possible models of extended producer responsibility for textiles. Producer responsibility models for textiles (ProTex). ISSN 1862-4804. In collaboration with Agnes Bünemann, Sabine Bartnik, Stephan Löhle and Nicole Kösegi. Ed. by Federal Environment Agency. Federal Environment Agency. Dessau-Roßlau (Texts 146/2023, research code 3722 33 305 0 FB001249), *https://www.umweltbundesamt.de/publikationen/erarbeitung-moeglicher-modelle-der-erweiterten*, last checked on 01.11.2023.

2.3 Circular Economy and Textiles in the EU

Gabriella Waibel, Policy Officer for Circular Economy at the European Commission, Nicole Espey, Institut für Textiltechnik of RWTH Aachen University

2.3.1 Toward a Circular Economy: Political Initiatives of the European Union

In the face of climate change and increasing environmental pollution, the world is experiencing considerable challenges. One significant factor for this development is unbridled economic growth at the expense of the environment. Our current linear production and consumption patterns ought to be decoupled from the increasing use of resources and be redirected toward environmentally friendly circular production processes. The European Union (EU) addressed these challenges through the adoption of the ambitious European Green Deal [1] and identified the transition to a circular economy as one pillar supporting a more sustainable Europe. Subsequently, the EU revised its Circular Economy Action Plan (CEAP) in 2020, aiming to make sustainable products the norm in the EU, with textiles among seven value chains identified as being key, due their negative environmental impact.

In this context, the EU's objective relates in particular to the rapid increase in clothing and textile consumption in recent years that is widely known as fast fashion and to the associated growth of textile waste. The European Commission's counter-concept for this development is summed up in the slogan "Fast Fashion is out of Fashion" and is specifically laid down in the EU Strategy for Sustainable and Circular Textiles, which sets out political and legal actions for the textile sector.

[This chapter was written in February 2024. As a result, some information, data, and references may not reflect the most current developments or updates. Readers are encouraged to verify details from more recent sources, where necessary.]

2.3.2 Textiles: A Sustainability Issue for the EU

Why textiles and garments?

Hardly any other product has experienced such a worldwide growth in consumption and simultaneous decline in value in recent years as apparel. Fast fashion is a synonym for the rapid production, use and disposal of garments that entered the global apparel market at the turn of the millennium and is now responsible for the growing mountain of textile waste worldwide.

---

1 European Commission, The European Green Deal, COM/2019/640 final. (2019)

According to a recent report by the European Environment Agency (2022), European households consume large amounts of textile products. In 2019, Europeans spent on average EUR 600 on clothing, EUR 150 on footwear and EUR 70 on household textiles. Average textile consumption per person amounted to 6.0 kg of clothing, 6.1 kg of household textiles and 2.7 kg of shoes in 2020. Calculations of the 'estimated consumption' based on production and trade data from 2020, and excluding industrial/technical textiles and carpets, show that total textile consumption is running at 15 kg per person per year. For 2020, this amounts to a total consumption of 6.6 million tons of textile products in Europe. Moreover, textiles are highly globalized, with Europe being a significant importer and exporter. In 2020, 8.7 million tons of finished textile products, with a value of EUR 125 billion, were imported into the EU-27.[2]

In this framework, "fast fashion" is an accumulation of different, very specific factors related to the following unique characteristics of the apparel sector. Despite the fact that there is no official definition of "fast fashion" in literature or on EU policy level, we would define it in our context as follows:

- The fast fashion business model is driven by vertically organized retail chains, where production and retail coincide within the same organizational structure. Therefore, the additional financial margin on the sale of an individual clothing item, which is normally charged by the retailer in the traditional distribution channel, is omitted. Thus, clothing can be sold more cost-effectively by vertical suppliers than in the traditional industry–retail distribution model.

- With the elimination of the retail distribution stage, this business model is also becoming faster from production to sale. Instead of launching 2–4 collections a year, like traditional fashion companies, fast fashion providers launch 12–24 collections a year.

- Normally, fast fashion companies employ large fashion design departments that constantly observe and imitate fashion shows and trends. New models are quickly brought onto the market, thus creating constantly new trends and further demand for new collections, which are continually being introduced.[3]

- Synthetic fibers, such as polyester and nylon, are the feedstock for about 60% of clothing and 70% of household textiles.[4] This raw material for all "fast fashion dreams" is based on crude oil. As long as oil prices are low, polyester can therefore be produced very cost-effectively and, correspondingly, synthetic apparel may also be offered in a low-price range.

---

[2] European Environment Agency: Textiles and the environment: the role of design in Europe circular economy, 2022, *https://www.eea.europa.eu/publications/textiles-and-the-environment-the*

[3] Vertica Bhardwaj and Ann Fairhurst., 2010: Fast fashion: response to changes in the fashion industry.

[4] European Environment Agency, 2021: Plastic in textiles: towards a circular economy for synthetic textiles in Europe

▪ The manufacturing costs of clothing are related to the highly economical labor costs in textile-producing countries in the Global South and are therefore low as well. The value of EU apparel imports originating from developing countries was €94.7 billion in 2022 and has increased at an average yearly rate of 7% since 2017.[5]

▪ Textiles are not yet integrated into the "Extended Producer Responsibility scheme" of the public waste management systems within the EU. Therefore, textile producers are not obliged to pay fees for disposing of the textiles. Accordingly, there is no financial incentive to produce less or to a rather more realistic market demand. The targeted revision of the Waste Framework Directive of July 2023, explained more in detail in section 3 below, aims to introduce Extended Producer Responsibility schemes in EU Member States. Overproduction is a significant issue in the fashion industry and generates enormous amounts of unused clothing articles that end up as waste.[6] Europeans use nearly 26 kilos of textiles and discard about 11 kilos of them every year.[7]

As a result of this fast-fashion trend, Europe faces major challenges in the management of used textiles, including textile waste. Between 2000 and 2019, the amounts of used textiles exported from the EU increased from a little over 550,000 tons to almost 1.7 million tons. As reuse and recycling capacities in Europe are limited, a large share of used textiles collected in the EU are traded and exported to Africa and Asia, where a significant amount of textile waste is likely to end up in landfill.[8] Because of this practice and its adverse effect on the environment and society, the European Commission has proposed an updated law on the shipment of waste, including rules governing the transport of waste – including textile products – across borders.

In summary: From a political perspective, due to the world-wide repercussions of fast fashion, apparel has become a waste issue. It is therefore a good thing, from the perspective of waste prevention, involving the rethinking of product design and development, collection, sorting and recycling, that the EU is currently addressing the sector with new legal regulations on circular economy.

It has also become clear that clothing and textiles have further major environmental impacts: from raw material sourcing, including the usage of pesticides, water and landmass for the cultivation of some fibers to the production processes related to spinning, dyeing and finishing and the use of hazardous chemicals and the enormous amount of

---

[5] Netherlands Ministry of Foreign Affairs, CBI (Centre for the Promotion of Imports from developing countries), 28 November 2023: "What is the demand for apparel on the European market". *https://www.cbi.eu/market-information/apparel/what-demand*

[6] European Environment Agency, 2019: Textiles and the Environment in a Circular Economy.

[7] European Parliament, 2023: The impact of textile production and waste on the environment, *https://www.europarl.europa.eu/news/en/headlines/society/20201208STO93327/the-impact-of-textile-production-and-waste-on-the-environment-infographics*

[8] European Environment Agency, 2022: EU export of used textiles in Europe's circular economy

process water. These impacts and more often than not occur in non-EU countries, where most textile production takes place.[9]

Furthermore, the greenhouse gas emissions arising from production facilities and the global transport of garments, as well as the whole consumption pattern, such as water, energy and chemicals used in washing, tumble-drying and ironing and finally the microplastics shedding into the environment, are having tremendous environmental effects.

Fact is, textiles have the fourth-highest impact on the environment and on climate change on average out of all categories of EU consumption relating to the environment and climate, following consumption of food, housing and mobility. Therefore, there is a need for political action.[10]

2.3.3 The EU Institutions

The European Union is led by four main decision-making institutions. These institutions collectively provide the EU with policy direction and play different roles in the law-making process:

- The European Council
- The European Commission
- The European Parliament
- The Council of the European Union

European Council:

The European Council brings together the heads of state or government of the EU member states, and defines the general political direction and priorities of the European Union to be implemented by the other three institutions.[11]

European Commission:

The European Commission is the EU's politically independent executive arm.

The EU Commission's main responsibilities include drawing up proposals for new European legislation for scrutiny and adoption by the Parliament and the Council of the European Union, managing most EU policies and the EU's budget, and ensuring that the EU member states apply EU law correctly.[12]

---

[9] European Environment Agency, 2019: Textiles and the Environment in a Circular Economy.

[10] European Environment Agency, 2023: Indepth topics, Textiles: *https://www.eea.europa.eu/en/topics/in-depth/textiles*

[11] European Council: *https://www.consilium.europa.eu/en/european-council/*

[12] European Commission: *https://european-union.europa.eu/institutions-law-budget/institutions-and-bodies/search-all-eu-institutions-and-bodies/european-commission_en*

European Parliament:

The European Parliament is composed of 705 members (MEPs) of seven political groups, who are directly elected every five years by the EU citizens. Together with the Council of the European Union, it is responsible for the adoption of legislation. The Parliament and Council have been compared with the two chambers of a bicameral legislature [52]. However, there are some differences from national legislatures; for example, neither the Parliament nor the Council has the power to initiate legislation. Therefore, while Parliament can amend and reject legislation, if it wants to make a proposal for legislation, it needs the Commission to draft a bill before anything can become law.[13]

Council of the European Union:

The Council of the European Union represents the member states' governments. It is where national ministers from each EU member state meet to adopt laws and coordinate policies. The Council of the European Union has the following main tasks: negotiation and adoption of EU laws, coordination of member states' policies, development of the EU's common foreign and security policy, conclusion of international agreements and the adoption of the EU budget. As for the Council's presidency, each member state takes turn at presiding over Council meetings for six months.[14]

2.3.3.1 EU: The Decision-Making Process

The main legal instruments of the EU are Directives and Regulations, enact EU legislation either "indirectly" or "directly."

▪ **Directive:**

This legally binding act of the European Union establishes a set of objectives which all member states of the European Union must fulfil. The member states are required to implement Directives into their own national legislation. They are free to choose the manner they see fit to fulfil the required objectives.[15] Directives are therefore more of an "indirect" type of legislation.

▪ **Regulation:**

This legally binding act of the European Union is directly applicable in all member states of the European Union. A Regulation is similar to national legislation in terms of the impact and direct effect it exerts. As such, a Regulation is the most pervasive of all the legal instruments of the EU.[16]

▪ **The Trilogue:**

---

[13] European Parliament: *https://www.europarl.europa.eu/portal/de*

[14] Council of the European Union: *https://www.consilium.europa.eu/en/*

[15] EU Monitor: "Directive": *https://www.eumonitor.eu/9353000/1/j9vvik7m1c3gyxp/vh75mdhkg4s0*

[16] EU Monitor: "Regulation": *https://www.eumonitor.eu/9353000/1/j9vvik7m1c3gyxp/vh75mdhkg4s0*

In the context of the European Union's ordinary legislative procedure, a trilogue is an informal inter-institutional negotiation that brings together representatives from the European Parliament, the Council of the European Union and the European Commission with the aim of reaching a provisional agreement on a legislative proposal that is acceptable to both the Parliament and the Council, the co-legislators. This provisional agreement must then be adopted by each of those institutions' formal procedures and then be implemented.[17]

2.3.3.2 EU: The Six Political Priorities (2019–2024)

Following the *European elections in May 2019*, the EU set a number of priorities to shape the political and policy agenda until 2024. They serve to address the main challenges faced by the EU and the member states. As the next EU Elections are going to take place in June 2024, we are, at the time of writing in spring 2024, not able to foresee all potential changes concerning the next political agenda. Therefore, we refer to the current set of political priorities (2019–2024) of the current European Union, under the guidance of Commission President Ursula Von der Leyen.[18]

A European Green Deal

Transforming the EU into a modern, resource-efficient and competitive economy, while preserving Europe's natural environment, tackling climate change and making Europe carbon-neutral and resource-efficient by 2050.

A Europe fit for the digital age

Embracing digital transformation by investing in businesses, research and innovation, reforming data protection, empowering people with the skills necessary for a new generation of technologies and designing rules to match.

An economy that works for people

Strengthening the EU economy while securing jobs and reducing inequalities, supporting businesses, deepening the economic and monetary union and completing the banking and capital markets union.

A stronger Europe in the world

Strengthening the EU's voice on the world stage by improving its standing as a champion of strong, open and fair trade, multi-lateralism and a rules-based global order. Boosting relations with neighboring countries and partners as well as strengthening the EU's ability to manage crises based on civilian and military capabilities.

Promoting our European way of life

---

[17] EUR Lex, Legal information system of the European Union: *https://eur-lex.europa.eu/EN/legal-content/glossary/trilogue.html*

[18] EU Commission: "The European Commission's priorities": *https://commission.europa.eu/strategy-and-policy/priorities-2019-2024_en*

Upholding fundamental rights and the rule of law as a bastion of equality, tolerance and social fairness. Addressing security risks, protecting and empowering consumers, as well as developing a system for legal and safe migration while effectively managing the EU's external borders, modernizing the EU's asylum system and cooperating closely with partner countries.

A new push for European democracy

Strengthening Europe's democratic processes by deepening relations with the European Parliament and national parliaments, protecting EU democracy from external interference, ensuring transparency and integrity throughout the legislative process, as well as engaging more widely with Europeans in shaping the EU's future.

The following section will explore some of the initiatives adopted within the framework of the European Green Deal.

2.3.3.3 The EU's Regulatory Framework for Textile Products

The focus on textiles of the CEAP materialized in the EU Strategy for Sustainable and Circular Textiles [3], a policy framework published in March 2022.

Traditionally, the textile industry has followed a linear model of production, consumption, and disposal. This model involves extracting raw materials, manufacturing products, and ultimately disposing of them after use. This linear approach results in substantial waste generation, resource depletion, and environmental degradation. Recognizing these challenges, the EU has set out to reshape the industry's dynamics.

The strategy focuses on the entire life cycle of textiles, from design and production to consumption and end-of-life treatment. By addressing environmental challenges, the strategy seeks to integrate sustainability into every aspect of the industry.

The EU recognizes the need for fair working conditions, ethical sourcing, and social responsibility within the textile industry with a view to ensuring a balanced approach that encompasses both environmental and social considerations for a truly sustainable and circular textiles sector. Initiatives aiming at increasing the sector's social sustainability are being proposed and implemented at EU level. They are referenced in the strategy, which, however, focuses on measures addressing the environmental footprint of the sector.

By introducing a set of initiatives outlined in the strategy, the EU aims to set a policy framework to jumpstart the textile industry's transition away from linear and unsustainable practices and toward an industrial ecosystem where "fast fashion is out of fashion," competitiveness is increased and a level playing field is established.

This chapter was written in February 2024, at a time when the legislative process of some of the initiatives proposed under the strategy was still ongoing. It means that the initiatives described in this chapter are still subject to change, depending on the outcomes of this process.

Section 2.3.3.3.1 to Section 2.3.3.3.4 outline some of the main initiatives included in the strategy.

2.3.3.3.1 Design Phase

The EU's Eco-Design for Sustainable Products Regulation [4] (ESPR) represents a ground-breaking step toward fostering sustainability in various industries, including textiles. Introduced as a stepping-stone of the CEAP, this regulation empowers the EU to set mandatory environmental performance and information requirements for products placed on the European market.

The regulation was adopted together with the EU Strategy for Sustainable and Circular Textiles in March 2022, and the co-legislators reached a provisional agreement on it in December 2023.

It expands the scope of the 2009 Eco-Design Directive [5], which introduced the approach of setting mandatory eco-design requirements for energy-using and energy-related products sold in the EU.

The proposed regulation will establish a framework for setting requirements through secondary legislation on a product-group basis [6]. The textiles product group has scored high in the prioritization exercise carried out by the European Joint Research Center (JRC) and co-legislators agreed to include it as a priority for the first Work Plan in the provisional compromise agreement. As a result, the JRC is already working on a preparatory study that will be supporting a future delegated act under ESPR.

Product aspects aiming to boost circularity that will be considered include:

- Product durability, reusability, upgradability and repairability
- Presence of chemical substances that inhibit reuse and recycling of materials
- Energy and resource efficiency
- Recycled content
- Carbon and environmental footprints
- Available product information, in particular a digital product passport

Based on those aspects, adequate eco-design requirements would be set, which manufacturers would be obligated to meet, fostering a shift toward more sustainable and circular practices in the textile sector.

Thus, the regulation is expected to drive innovation in textile production, including through the use of recycled or sustainably sourced materials, efficient manufacturing processes and designs that prioritize longevity and recyclability.

The regulation proposes mandatory green public procurement (GPP). This would compel public authorities to consider ecological factors when purchasing goods, ensuring that sustainability becomes a fundamental criterion in procurement decisions. By mandating green criteria, for instance energy efficiency and environmental impact, the regulation incentivizes the adoption of eco-friendly practices along the supply chain. GPP criteria already exist at EU level [7] for textile products and others, but they have only been voluntary.

The ESPR also contains measures for ending the wasteful and environmentally harmful practice of destroying unsold consumer products. Companies will have to take measures to prevent this practice and the co-legislators have introduced a direct ban on destruction of unsold textiles and footwear products, with derogations for small companies and a transition period for medium-sized ones. In addition, large companies will need to disclose every year how many unsold consumer products they discard and why. This is expected to strongly disincentivize businesses from engaging in this practice.

2.3.3.3.2 Tackling Greenwashing

In addition to the measures mentioned above, the ESPR will introduce information requirements, which will likely feed into a digital product passport (DPP). The introduction of a DPP for a set of product groups, including textiles, reflects a progressive approach toward fostering environmental sustainability in the digital realm.

This innovative measure aims to enhance transparency and traceability throughout a product's life cycle by creating a comprehensive digital profile – the "passport" – encompassing key environmental and social impact data. By offering consumers, manufacturers and regulators easy access to information, such as a product's energy efficiency, recyclability, and supply chain details, the initiative empowers stakeholders to make informed choices that align with ecological considerations.

This digital initiative not only facilitates compliance with evolving sustainability standards, but also encourages a shift toward responsible consumption and production practices, fostering a holistic approach to environmental stewardship in the digital product landscape.

Furthermore, in March 2023, the European Commission adopted a proposal for a Green Claims Directive[19], which represents a significant stride toward enhancing transparency and accuracy in environmental labeling. This has profound implications for the textile industry. This Directive, part of the broader sustainable product policy framework, addresses greenwashing concerns by establishing clear guidelines for eco-friendly claims on products, including textiles. Manufacturers will now be required to substantiate their environmental claims with third-party evidence. This will promote integrity in marketing practices and improve the functioning of the internal market for businesses engaging in truly sustainable practices.

For the textile industry, this Directive imposes a notable shift toward sustainability and responsible production. Companies engaged in textile manufacturing must provide verifiable information regarding their products' environmental performance,

---

[19] *https://environment.ec.europa.eu/publications/proposal-directive-green-claims_en*

covering aspects such as resource use, chemical composition, and overall environmental impact. This ensures that consumers can make informed choices and will
foster a marketplace where genuine eco-friendly products are distinguished from
misleading alternatives.

The Green Claims Directive is poised to stimulate innovation within the textile sector,
prompting the adoption of sustainable materials, processes and supply chain practices.
Textile businesses will need to align their operations with stringent environmental
standards; this will potentially lead to increased investment in eco-friendly technologies and practices. It complements the Empowering Consumers in the Green Transition initiative in its revision [8] of the Unfair Commercial Practices Directive, which
was finally adopted by the co-legislators in early 2024.

Following this revision, certain commercial practices such as making generic environmental claims and making false claims about the durability and the repairability of a
product will be banned. Striking a balance between the interests of businesses and
consumers, the updated framework promotes ethical conduct, aligning with contemporary standards of corporate responsibility. Through this initiative and the Green Claims
Directive, the EU aims to strengthen consumer trust, ultimately fostering a more equitable and competitive economic environment.

In parallel, the Commission is aligning its EU Ecolabel criteria for textile products [9]
with the work done under ESPR. The EU Ecolabel is the European Union's official label
that rewards best-in-class products and services for environmental excellence. This
label guarantees adherence to stringent eco-friendly standards throughout the product's life cycle – encompassing fiber production, the manufacturing process and a
strict restriction on the use of hazardous substances. The label also indicates compliance with social criteria and criteria that guarantee high quality and durable products.

2.3.3.3.3 Waste Actions

A circular economy is characterized by a closed loop, where the value of products, materials and resources is maintained in the economy for as long as possible and waste
generation is minimized. Waste is managed according to the waste hierarchy, which
prioritizes circular and more sustainable practices such as reduce, reuse, recycle, then
recovery and, ultimately, disposal.

The upcoming obligation to establish separate waste collection for textiles in the EU
reflects a significant step toward a more circular and sustainable economy. Envisioned
under the 2018 revision of the Waste Framework Directive [10], it mandates EU member states to implement systems for ensuring the separate collection of used textiles by
1 January 2025. The move is pivotal in addressing the environmental impact of textile
waste, fostering recycling, and reducing the overall environmental footprint of the industry. By separately collecting textiles, the EU aims to enhance the efficiency of recycling processes so as to promote the re-utilization of materials and mitigate the envi-

ronmental consequences associated with textile disposal. This initiative not only aligns with broader EU sustainability goals, but also encourages a paradigm shift in consumer behavior, emphasizing responsible consumption and contributing to a more resource-efficient and environmentally aware future.

In July 2023, the European Commission published a proposal for a targeted revision of the Waste Framework Directive [11] that focused on food and textiles. Its main feature is the introduction of mandatory Extended Producer Responsibility (EPR) schemes for textiles at member state level in the EU. This would hold producers responsible for the environmental impact of their products, from production to end of life. Mandatory EPR schemes would contribute significantly to increasing circularity by incentivizing producers to design products with sustainability in mind. This includes using recyclable materials, reducing waste generation and promoting easy disassembly for recycling. Producer responsibility organizations are expected to establish systems for the separate collection, reuse, preparation for reuse, recycling, and proper disposal of textile products they make available on the market for the first time within the territory of an EU member state.

Textiles products' lives do not end once they become waste. While intra-EU trade of waste is very common, a big part of textile waste (1.7 million tons in 2019 [12]) is actually exported from the EU to third countries, oftentimes under the guise of second-hand goods. European co-legislators reached a provisional agreement on a new Waste Shipment Regulation [13] in November 2023, which introduces stringent measures to control the shipment of textile waste. The regulation aims to facilitate shipments in line with the objectives of the circular economy, while strengthening the rules for waste shipment within the EU and to non-EU countries, both OECD and non-OECD. These measures will not only reduce the risk of environmental harm linked to textile waste, but also promote transparency and accountability and foster sustainable practices in the management of such waste.

2.3.3.3.4 Enabling Conditions

While the transition of the EU's textiles sector toward a circular economy is widely based on a regulatory framework, this is accompanied by strong enabling conditions that make the policy implementation smooth and swift.

These enabling conditions include the ongoing support to research and innovate in the sector, for instance through funding programs such as LIFE, Horizon Europe and Recovery plans, which are needed for the development and implementation of new circular technologies and processes.

Given that 2023 and 2024 were designated the European Years of Skills, significant attention has been placed on the need for new skills and re-skilling in the textile industry. This emphasis is particularly relevant for small and medium-sized enterprises (SMEs), which represent 99% of businesses operating in the EU textiles sector. Among others, the EU launched a Textiles, Clothes, Leather and Footwear (TCLF) Skills Alliance [14] under the EU Pact for Skills, in order to increase the upskilling and re-skilling of the European workforce and the acquisition and transfer of green and digital skills.

A Transition Pathway for the Textiles Ecosystem [15] has been put in place through a co-creation process which saw strong engagement on the part of businesses. A policy document emerged from this exercise in 2023, which was followed by a call for pledges from industry to gain the sector's commitments to sustainability and circularity.

2.3.4 Conclusion

European efforts toward enabling a circular economy in the textiles ecosystem are a beacon of progressive policy-making and demonstrate a bold commitment to sustainability and environmental stewardship. Through a comprehensive framework of legislation, initiatives and incentives, the European Union has embarked on a transformative journey toward a circular economy model within the textile industry. By prioritizing resource efficiency, waste reduction, and eco-design principles, these measures not only address pressing environmental concerns, but also pave the way for a more resilient and competitive economy.

Importantly, the significance of these measures extends far beyond the borders of the European Union. As a global leader in sustainability practices, the EU is setting a precedent for other regions and nations to follow, fostering international cooperation and collective action toward a more sustainable future for the textile industry. Furthermore, these measures have the potential to catalyze innovation and stimulate economic growth and create new opportunities for businesses, entrepreneurs, and consumers alike.

In essence, the European measures on textiles and circular economy underscore the imperative of integrating environmental considerations into policy-making and industry practices. By embracing the principles of circularity and a holistic approach to sustainability, Europe is charting a course toward a more prosperous, equitable and environmentally sound future for generations to come.

References for Section 2.3

[1] European Commission, *The European Green Deal*, COM/2019/640 final. (2019)

[2] European Commission, *A new Circular Economy Action Plan for a cleaner and more competitive Europe*, COM (2020) 98 final. (2020)

[3] European Commission, EU Strategy for Sustainable and Circular Textiles, COM/2022/141 final. (2022)

[4] European Commission, Proposal for a *REGULATION OF THE EUROPEAN PARLIAMENT AND OF THE COUNCIL* establishing a framework for setting eco-design requirements for sustainable products and repealing Directive 2009/125/EC, 2022/0095 (COD). (2022)

[5] European Commission, *DIRECTIVE 2009/125/EC OF THE EUROPEAN PARLIAMENT AND OF THE COUNCIL of 21 October 2009* establishing a framework for the setting of eco-design requirements for energy-related products. (2009)

[6] European Commission, *New product priorities for Eco-design for Sustainable Products*. Accessed 29 January 2024, *https://ec.europa.eu/info/law/better-regulation/have-your-say/initiatives/13682-New-product-priorities-for-Ecodesign-for-Sustainable-Products/public-consultation_en*

[7] European Commission, *Green Public Procurement Criteria and Requirements*. Accessed 29 January 2024, *https://green-business.ec.europa.eu/green-public-procurement/gpp-criteria-and-requirements_en*

[8] European Commission, Proposal for a *DIRECTIVE OF THE EUROPEAN PARLIAMENT AND OF THE COUNCIL* amending Directives 2005/29/EC and 2011/83/EU as regards empowering consumers for the green transition through better protection against unfair practices and better information, COM/2022/143 final. (2022)

[9] European Commission, 2014/350/EU: *Commission Decision of 5 June 2014 establishing the ecological criteria for the award of the EU Ecolabel for textile products*. (2014)

[10] European Commission, *Directive (EU) 2018/851 of the European Parliament and of the Council of 30 May 2018* amending Directive 2008/98/EC on waste. (2018)

[11] European Commission, Proposal for a *DIRECTIVE OF THE EUROPEAN PARLIAMENT AND OF THE COUNCIL* amending Directive 2008/98/EC on waste, COM (2023) 420 final. (2023)

[12] European Environment Agency, *EU exports of used textiles in Europe's circular economy*. (2023)

[13] European Commission, Proposal for a *REGULATION OF THE EUROPEAN PARLIAMENT AND OF THE COUNCIL* on shipments of waste and amending Regulations (EU) No 1257/2013 and (EU) No 2020/1056, COM(2021) 709 final. (2021)

[14] European Commission, *Pact for Skills – Textiles ecosystem and LSP(s)*. Last accessed 29 January 2024, *https://pact-for-skills.ec.europa.eu/about/industrial-ecosystems-and-partnerships/textiles_en*

[15] European Commission, *Internal Market, Industry, Entrepreneurship and SMEs – Textiles Ecosystem Transition Pathway – Co-creation and co-implementation process*. Last accessed 29 January 2024, *https://single-market-economy.ec.europa.eu/sectors/textiles-ecosystem/textiles-transition-pathway_en*

2.4 Progress in Circular Innovation Practice of the Textile Industry in China

Yan Yan, Office for Social Responsibility of China National Textile and Apparel Council

2.4.1 The Global Consensus Now is to Develop a Circular Economy

The linear economic development model developed after mankind entered the industrial age is becoming increasingly unsustainable, as more and more prominent disadvantages emerge. To name a few, such development is being accompanied by a lack of non-renewable resources, deterioration of the ecological environment, climate change and so on, posing a major threat to human survival and development. Compared with this linear model, the circular economic development model aims to achieve efficient resource utilization through reducing inputs at source, improving resource efficiency in the production process, and recycling waste at the end of the product life cycle. At present, major economies of the world generally regard the development of circular economy as the basic path for solving the resource and environmental constraints, tackling climate change and fostering new drivers of growth. Many developed countries and regions, including the United States, the EU and Japan, have made systematic arrangements for the implementation of a new round of circular economy action plans, accelerating the development of circular economy and tackling new global resource and environmental challenges. As a key component of global economy, the fashion industry is recognized as the "seventh largest economy in the world." So, promoting the circular development of the fashion industry holds great significance for sustainable development across the entire world.

In this context, to develop circular economy is the only way for sustainable development of the global fashion industry. China is an important textile and garment producer, exporter and consumer in the world, and China's textile and garment industry is a key part of the global fashion industry. In 2018, China's total fiber processing accounted for 50% of the world's total and, in 2022, China's textile and garment exports accounted for 36% of the world's total. On the other hand, over the past decade, China's market share of global fashion industry has grown to 38%, ranking among the top three markets for many world-renowned fashion brands and enterprises. The key to circular development of the global fashion industry lies in China's textile and garment industry.

Against such backdrop, China has incorporated the development of circular economy into major national strategies, and taken it as an effective way to resolve the dual constraints of resources and environment, cope with climate change, and promote economic and social transformation and development. To adapt to the development trend and implement the national development strategy, China National Textile and Apparel Council (CNTAC) has taken the initiative to lead and promote the circular transformation and development of China's textile and garment industry, and carried out a series of industry work, producing fruitful results.

Taking the project "Circular Fashion: China's New Textile Economic Outlook" as an example, in order to gain an in-depth understanding of the circular development status of, and prospects for, China's textile and garment industry, CNTAC has, in conjunction with the Ellen MacArthur Foundation and the company Lenzing AG, Austria, carried out a series of investigations and studies. After a year of dedicated effort, the research report is now complete. The key findings are presented in sections 2.4.2, 2.4.3, 2.4.4 and 2.4.5.

2.4.2 Key Findings I

China keeps improving its policies on the development of circular economy, playing an important role in supporting and promoting the historical process of circular transformation and development of the textile and garment industry.

Since 2005, China has attached great importance to the development of circular economy and incorporated circular economy into national development strategies. The state has, by constantly improving the policy system and building a sound institutional system of circular economy, comprehensively and thoroughly promoted the circular transformation and development of the economy and society with sustained efforts and produced remarkable results in all areas. The development of circular economy is transitioning from a "following stage" to a "leading stage."

Let us look at it stage by stage: In 2005–2012, the development of circular economy in China was in its initial stage on the whole, during which a preliminary circular economy policy system was established. At that time, the policy concept changed from "economic development" to "circular development." The policy focus transitioned from primarily emphasizing the production end to encompassing the entire spectrum of production, construction, circulation and consumption. Accordingly, China started to take comprehensive policy measures instead of individual measures in specific sectors of the economy During this period, guided and promoted by the circular economy policies, China's textile industry achieved significant reductions in indicators such as energy consumption per unit of industrial added value and water consumption per unit of industrial added value. Progress was made in exploring circular production, utilization and other aspects.

At this stage, China issued the following main policy documents on the development of circular economy:

- In 2005, the *Several Opinions on Accelerating the Development of Circular Economy* was released. This was the first strategic plan issued by China for developing circular economy. Since then, the development of circular economy has been formally incorporated into the state's economic and social development plan. The Opinions explicitly require key industries, such as textiles, to intensify efforts in the management of resource consumption, such as energy, raw materials and water in key industries like textiles, so as to reduce consumption and improve resource utilization.

- In 2006, the *Outline of the Eleventh Five-Year Plan for National Economic and Social Development* was released, incorporating the development of circular economy into the five-year development plan in a dedicated chapter. It was made clear in the document that the textile industry should intensify efforts in the development of circular economy and increase the industry's added value.

- In 2008, the *Circular Economy Promotion Law of the People's Republic of China* was released. It was the first legal document issued by China on circular economy and made China the third country in the world to promote circular economy through legislation.

- In 2011, the *Outline of the Twelfth Five-Year Plan for National Economic and Social Development* was released. Different from the Eleventh Five-Year Plan, this document is characterized by more comprehensive and in-depth elaboration and arrangements for the development of circular economy.

From 2013 to 2020, China promoted the development of circular economy extensively in various industries and areas, and proved to be highly successful in this regard. At the same time, China's policy system for circular economy development basically took shape, covering a wide range of comprehensive areas and developing more abundant policy tools. These policies attached greater importance to segmented areas, among which waste resources such as waste textiles became the focus of circular economy policies. At this stage, China basically established a circular industrial system for its textile industry, and saw rapid development of its waste textile recycling industry as well.

At this stage, China issued the following main policy documents on the development of circular economy:

- In 2013, the *Circular Economy Development Strategy and Near-Term Action Plan* was released, providing a comprehensive plan for the development of circular economy, setting out the development goals in the short, medium and long terms, and identifying key areas, key projects and major projects for the development of circular economy. Regarding the development of a recycling industrial system, it clarified the development goals and key tasks in key areas such as the textile industry.

- In 2015, China released the *Opinions on Accelerating the Construction of Ecological Civilization,* incorporating the development of circular economy into the overall plan for ecological civilization progress and further highlighting the strategic position of circular economy. It explicitly encourages the recycling and utilization of waste items, such as textiles and auto tires.

- In 2016, the *Outline of the 13th Five-Year Plan for National Economic and Social Development* was released, further highlighting the circular link between production and life as well as the utilization of waste as a resource, including industrial solid waste, construction waste and waste textiles, for the development of circular economy.

- In 2017, the *Circular Development Leading Action* was issued, signifying an upgrade to China's environmental management. The focus moved from end treatment with harmlessness at its core to an ecological cycle centered on resources and ecology. This transition is expected to promote rapid development of the circular economy industry.

- In 2018, China introduced the *Work Plan for the Zero-Waste City Pilot Project,* becoming the first country to unveil the concept of a "Zero-Waste City". This explores effective pathways for urban sustainable development by combining the circular economy with urban development and management.

In 2021, China ushered in a new stage of development, issuing a series of circular economy policies based on basic national conditions and reflecting the changing international situation. This provided for comprehensive and thorough arrangements for the development of circular economy and further boosted the development of circular economy. The waste textile recycling industry has become a key area for the circular development of China's textile industry, playing a significant role in in advancing circular economy.

Since 2021, China has issued the following main policy documents on the development of circular economy:

- In 2021, the Guiding Opinions on Accelerating the Establishment and Improvement of a Green and Low-Carbon Circular Development Economic System was released. This was the first instance of China implementing top-level design and overall arrangements for the establishment and improvement of a green, low-carbon and circular economic system. It underscores the state's increased emphasis on the coordinated promotion of economic development, eco-environmental protection and GHG emission control, which has a far-reaching impact on China's green development.

- In 2021, the Outline of the 14th Five-Year Plan (2021–2025) for National Economic and Social Development and the Long-Range Objectives Through the Year 2035 was released, proposing to promote waste classification, reduction and recycling and speed up the construction of a waste materials recycling system. It put forward new and stricter requirements for work related to circular economy.

- In 2021, China issued the *14th Five-Year Plan for Circular Economy Development*, proposing "Three Tasks, Five Projects, Six Actions and Four Safeguard Policies" with the goal of "raising the efficiency of resource utilization across the board" at its core and focusing on the reuse and recycling. The release of this document was highly significant for accelerating the development of a green low-carbon circular economy system and achieving the goals of "carbon peaking and carbon neutrality."

- In 2022, the *Implementation Opinions on Accelerating the Recycling of Waste Textiles* was released. This was the first special development plan issued by China on the recycling of waste textiles. As a long-term strategic plan, it clarified the key tasks of constructing the recycling system of waste textiles from the aspects of production, recovery, comprehensive utilization, etc., and accelerated the development process of the waste textile recycling industry in China.

2.4.3 Key Findings II

The basic model for circular economy development of China's textile and garment industry has begun to take shape.

In 2013, the Circular Economy Development Strategy and Near-Term Action Plan was officially released and implemented, setting out a comprehensive plan for the development of circular economy. It defined the development goals in the short, medium and long terms, and identified key areas, key projects and major projects for the development of circular economy. With respect to the construction of a circular industrial system, it clearly defined the development goals of the textile industry and proposed the key tasks, including raw material substitution, energy saving and resource recycling in production, standardization of waste textile recycling and construction of a circular economy chain for the textile industry. It is worth noting that this document, based on the practical experience and typical practices of circular transformation and development in China's textile and garment industry, introduced the basic model for the development of circular economy. This is of significant practical importance in promoting the circular development of the industry.

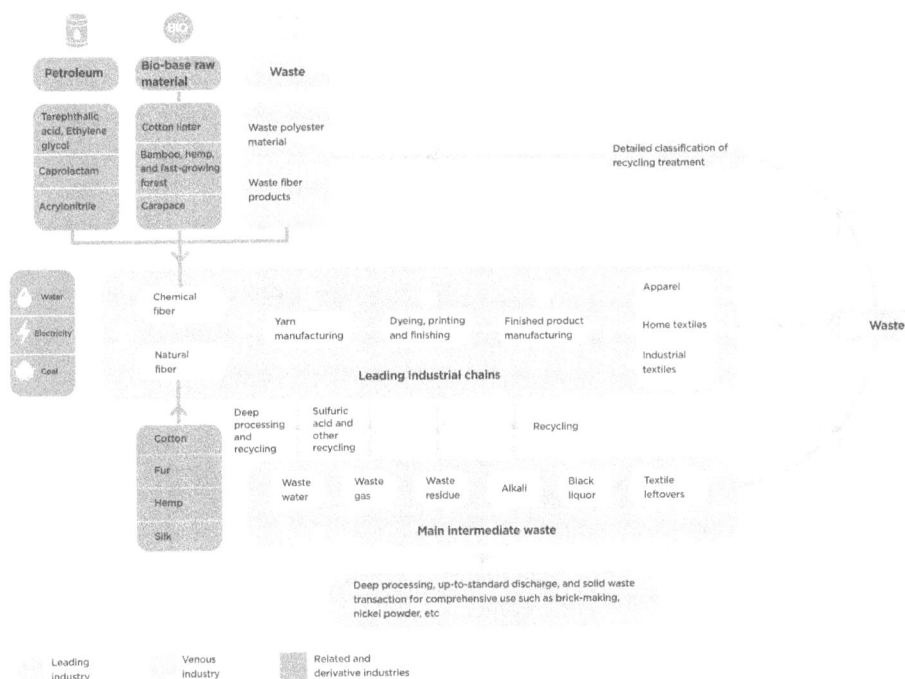

Figure 2.3 Diagram of basic model for circular economy development of the textile industry

2.4.4 Key Findings III

Substantial progress has been made in the circular development of China's textile and garment industry.

Firstly, the structure of fiber raw materials tends to diversify. The proportion of recycled fiber processing in China keeps increasing, from 9.6% in 2010 to 11.3% in 2015. According to Chinese estimates, in 2018, the output of recycled fibers in China exceeded 7 million tons, with recycled fibers being widely used in clothing, home textiles and other fields. Industrialization of bio-based chemical fibers has been achieved, with an annual production capacity of approximately 300,000 tons and an expanding variety of products. **Secondly, the level of resource recycling in the manufacturing process has been raised.** During the "12th Five-Year Plan" period, a large number of new technologies for energy saving, consumption and emission reduction were widely adopted. The fresh water intake for every 100 meters of printed cloth decreased from 2.5 tons to less than 1.8 tons and the water reuse rate increased from 15% to more than 30%. The obligatory targets for reducing energy consumption per unit of value added, water intake and total pollutants were met in full, and specifically, the energy consumption per unit of industrial value added, water consumption and emission of major pollutants were reduced by 20%, 30% and 10%, respectively, from those during the "11th Five-Year Plan" period. **Thirdly, we have basically established an industrial chain of comprehensive utilization of waste textiles.** At present, China has basically completed the construction of an industrial chain covering waste textile recycling, sorting and comprehensive utilization, established a diversified recycling system and developed some resource utilization technologies. As a result, scaled circular industries have emerged, such as, "from textile products to discarded textiles to textiles," "from textile products to discarded textiles to insulation materials," along with the rise of many leading demonstration enterprises. The level of industrial standardization and cleanliness has continued to improve. In 2018, the proportion of textile waste recycling reached 17%.

2.4.5 Key Findings IV

Emerging trends in green design and green consumption are fostering new developments that support circular development.

First, more explorations have been made in green design practice. Green design provides strong support for circular economy. Research in China shows that 80% of resource consumption and environmental impact depend on the product design phase. The research surveyed 40 companies in the textile and apparel industry, and the results revealed that 25 of these product design companies factor ecological design into the product design phase. Specifically, 24% of the companies factor ecological design into all their products, while 52% factor it into the majority of their products. The research surveyed 50 designers. It found that about 58% of the respondents do their best to apply green design, 7.8% of the respondents use green design in all of their designs, and 13% of the respondents seldom or never consider this issue. The conclusion reached is that green design has gradually become a key focus for companies and designers in product development and design.

In essence, most designers consider adopting green design because they believe green products represent the future of the consumer market and they agree with the concept of green development. In practice, designers focus on the use of green materials and the extension of product life cycle. The survey results show that green materials mainly include recycling materials and bio-based materials, of which bio-based cellulose fibers account for about 60.78%, organic materials account for about 52.94%, recycling materials account for about 54.9%, and transformed materials from recycling of waste clothes account for about 45.1%. In terms of extending the product life cycle, 58.82% of designers focus on extending the product life cycle by selecting appropriate materials and manufacturing methods. Additionally, 54.9% of designers increase the emotional connection with consumers by incorporating cultural, creative and other elements, thereby extending the product's usage time by consumers.

Second, trends in sustainable consumption are starting to emerge. Some consumers prefer green textile and clothing consumption, their awareness of sustainable consumption has been raised and most of them are willing to take action. Consumers' awareness of sustainable consumption has been raised. More than 67% of consumers focus on sustainable products and about 26% of consumers report that they not only focus on sustainable products, but also buy them; about 30% of consumers have an awareness of sustainable consumption, but do not know exactly what sustainable products are and how to buy them; less than 2% of consumers are not interested in sustainable products. Most consumers are willing to accept a green premium on a product, but the sustainability of a product is not the primary factor governing consumer behavior. More than 90% of consumers are willing to accept a premium for sustainable products, but quality is still the first factor considered by consumers in their purchase of clothes, followed by beauty, price, performance, and then sustainability. Consumers' openness to sustainable consumption needs to be further enhanced. It is worth noting that, although a high percentage of consumers are concerned about sustainable features and willing to accept a premium (67% and 90%, respectively), the consumers who bought specifically only accounted for 27%. This suggests a big gap between consumers' consumption consciousness and consumption behavior transformation in practice. It is important that the brands and enterprises take into account clothing quality, fashion and comfort when designing, producing and selling green clothing products, and set reasonable prices to guide consumer behavior in practice. Emerging new models and new forms of business such as clothing rental, sharing and resale have boosted sustainable consumption. The survey results show that 34% of consumers are willing to buy second-hand clothing mainly for wearing on special occasions, where there is no need to buy new products. Furthermore, the reasons for choosing the limited edition are as follows: it has collection value, price concessions and a unique design that highlights the personality; 72% of the consumers choose to rent clothing also for specific occasions, such as wedding dresses and performance clothing; on the emerging shared clothing rental platform, more than 53% of the consumers rent the same clothing categories as above, including special-occasion clothing, daily clothing, children's clothing, etc.

2.4.6 Remarkable Progress in Circular Transformation: Innovative Practices in China's Textile Industry

Overall, remarkable progress has been made in circular transformation of China's textile and garment industry, with a number of emerging innovative practice cases. However, we still have a long way to go for the circular development of the industry, due to such factors as the large scale and volume of the industry, long industrial chain, and diverse product categories. There is still much room for improving the substitution scale of non-renewable raw materials, the resource utilization level of manufacturing links, the recycling of waste textiles and so on, and there are still many challenges in terms of development policy, green technology, green design, professionals, and industrial collaboration. To be specific, the innovation in advanced technology and process equipment needs to be further enhanced; the policy system and support system underpinning the circular development of the industry need to be further improved; circular design as the source solution is still in the stage of concept advocacy; the driving force of the green consumption market for the industry is insufficient; a unified vision and synergy have not formed for the transformation of the industrial collaborative system; and there are insufficient professionals to promote circular fashion.

Vision is the soul and power source of industry development. The circular development of China's textile and apparel industry must have a unified vision. CNTAC has, based on a deep understanding of the circular pattern, the development characteristics and direction of China's textile and garment industry, and according to the circular development vision and trends in the international fashion industry, unveiled the vision for the circular development of China's textile and garment industry. This includes the tasks of building a green, circular and low-carbon textile and garment industry system, improving resource utilization efficiency, reducing resource consumption and negative environmental impact, and increasing the effective market supply of circular fashion clothing. In order to make the vision come true, CNTAC has proposed five goals for the circular development of China's textile and garment industry. These are: further optimize the structure of raw materials, reduce the consumption of non-renewable resources, transition to textile design based on the principle of circular economy, further improve resource efficiency in manufacturing processes, expand green consumption and improve the recycling and upgrading of waste textiles.

In further response to the challenges facing circular development of the industry and to explore effective solutions, CNTAC launched the "China Textile Recycling Innovation Research Project" in May 2022, which was aimed at conducting in-depth research on the status quo and issues in the recycling, sorting and reuse process of textiles, and identifying the issues and innovation opportunities for China's textile industry in its transition to the textile closed loop (T2T). In order to advance the project, CNTAC has invited major recycling manufacturers in the textile industry chain to participate jointly. We will prepare research reports and application guidelines on the basis of investigation and research, carry out capacity building and organize research results release activities for the participating manufacturers, showcase the industrial panorama, technology pathways and innovation trends in the recycling of major textile materials, and provide a reference for participants to identify investment, business and innovation opportunities in T2T supply chain collaboration.

2.4.7 Conclusion

Circular economy is of great significance to the sustainable development of the textile and garment industry. Over the years, CNTAC has prioritized the circular economy in its sustainable development agenda, conducting extensive explorations, achieving some accomplishments. However, circular transformation of the textile and garment industry represents a profound change. Promoting the industry's circular development is a comprehensive undertaking that requires collaborative efforts from various stakeholders, including government bodies, businesses, designers, research institutions, associations and consumers.

3

From Used Textiles to New Products: Innovations for the Circular Economy

3.1 Textile Value Chain Circularity and New Products

Uday Gill, Indorama Ventures Group (IVL)

3.1.1 Who Will Address the Unmanaged Textile and Plastic Waste?

In 2023, global oil production amounted to about 4.5 billion tons of crude oil. Oil consumption in the same year comprised about 45% to transportation, 42% in heating and energy, 8% in petrochemicals and 5% in others. The petrochemicals contribution can be further divided into construction (16%), packaging (36%), textile (15%) and other (33%). Disposal of petrochemicals comprised landfill (40–45%), leak out (27–35%), recycling (10–15%) and incineration (12–14%) (Figure 3.1). The share of recycling in total disposal is very small and most of it goes to landfill or escapes into the environment.

Plastic was invented as a substitute for metals, coal, and other industrial feedstocks. While plastic is a miracle of science, irresponsible use and disposal by humans has turned this solution into a problem.

Smart and sustainable solutions are evolving very rapidly, with carbon neutrality and biodiversity becoming new focus areas for businesses to address. Sustainability focus will help build new platforms for success and technology will help speed up these solutions, turning this challenge into an opportunity.

The world recognizes this and is gearing up to support and finance such activities. It is encouraging to see that several leading countries and corporates have committed to carbon neutrality over the next few decades. However, to begin reversing the damage, governments, civic society and businesses must collaborate and integrate on a global scale with the aim of achieving negative greenhouse gas emissions.

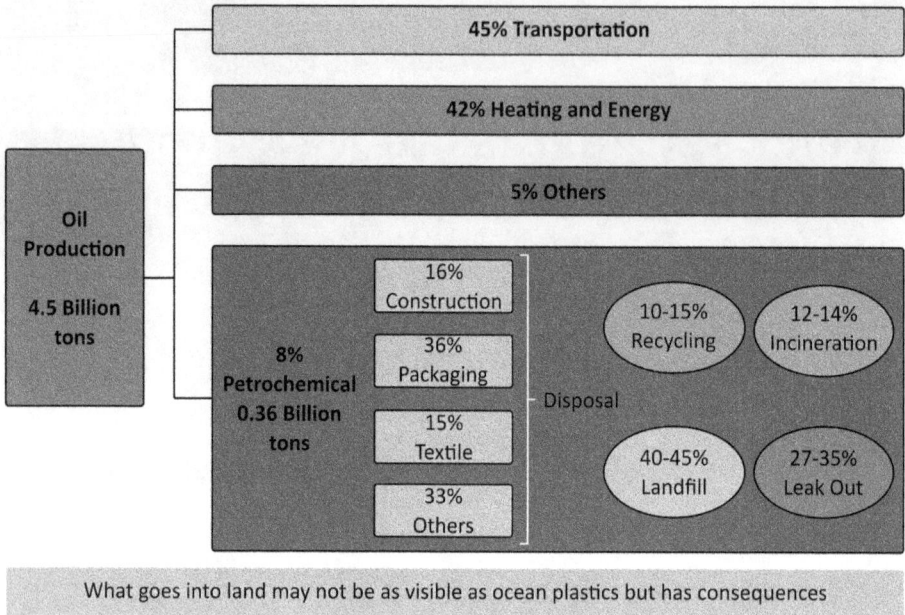

Figure 3.1 Overview of the percentage oil consumption per sector and disposal strategies for petrochemical products [1, 2, 3]

Modern consumers are aware of the consequences of climate change and are asking questions about our lifestyle. They have also started to take the environmental impact into account when making purchasing decisions.

Governments and industry have been making commitments to climate change reversal, which is a positive step. However, we need to do more by proactively informing the world through the disclosure of our actions and adherence to compliance measures.

3.1.2 Is Bottle-to-Textile the Answer to Circularity?

The textile industry has been using recycled bottles as feedstock for recycled textile products, accounting for 11% of its total consumption, for a while. That said, like-to-like textile recycling, the aspiration of the circular economy, has only reached 0.5%. As the bottle industry seeks to fulfill its own commitment for recycling to align with regulatory requirements and consumer expectations, the textile industry will in turn have to develop its own renewable and 'like-to-like' feedstock supply chain.

Figure 3.2 shows the material balance for used PET bottles for recycling activities in the textile and beverage industry, based on the commitments made by both sectors and on what will be left for textile applications once the beverage industry fulfills its own 'like-to-like' obligations.

Global Brands – Recycle Content Target	
The Coca Cola Company	25%
Pepsico	25%
Nestle	30%
Danone	50%
Mondelez	5%
Diageo	40%
Henkel	30%
Colgate-Palmolive	25%
Mars	30%
L'Oreal	50%
SC Johnson	15%
Kellog	30%
Essity	25%
FrieslandCampina	10%

Material Balance for Recycled PE and Used Textiles	2021	2025	2030
Polyester fibers consumption	57.0	66.7	76.4
PET Consumption (4% growth)	23.9	29.0	32.0
Collection rate	52%	54%	56%
Collected Volume	12.4	14.5	17.9
Yield	79%	79%	80%
Recycle flake production	9.8	11.5	14.3
Flake consumption			
Bottles 28% by 2025 and 35% by 2030	1.7	7.8	11.2
Industrial applications (Growth 4%)	2.0	2.3	2.9
Balance flakes for fibres	6.1	1.4	0.2
Industrial waste 2% for fibers	1.8	1.9	2.2
Demand for recycled polyester	7.7	20.0	38.2
Deficit assuming textile industry use 30% of recycle content by 2025, 50% by 2030		-16.7	-35.8

Recycling Target for Textile Brands	
Zara	2025: 100% organic linen and recycled polyester
Adidas	2024: 100% recycled polyester
H&M	2025/2030: 30% recycled material / 100% recycled materials
UNIQLO	2030: 50% recycled materials
Nike	2025: 100% waste deverted from landfill with at least 80% recycled back into products
Puma	2025: 75% recycled materials
Wrangler	2025: 100% sustainable cotton
BOSS	2025: 100% sustainably cooton and > 50% recycled synthetic fibers
GAP	2025: 60 % recycled polyester

Figure 3.2 The material balance of recycled PET and future recycling content targets of global and textile brands [4]

In the packaging sector, the EU introduced a policy targeting single-use plastics, aiming to collect 77% of beverage bottles by 2025 and 90% by 2029. Regarding recycled content within PET beverage bottles, EU targets are 25% by 2025 and 30% by 2030. The European Plastics Pact has imposed a plastics tax of 800 euro/MT that is forcing the packaging industry to increase recycled content in their products.

Beverage and non-beverage global brands have made a commitment that 28% of their consumption will consist of recycled PET by 2025. This is likely to reach 35% by 2030. As the demand for bottle-to-bottle recycling increases, the premium on flake price becomes unviable for textile use. To date, the textile industry has had ready access to bottle flakes as a feedstock, since countries like China, India, and Thailand have not yet permitted bottle-to-bottle recycling. Once these countries allow bottle-to-bottle recycling, there will be limited polyester bottle flakes available for textile use.

Therefore, the textile industry absolutely must design products for circularity, adopt new technologies, new business models, and realign supply chains to prepare for like-to-like recycling.

3.1.3 Are Bioplastics a Sustainable Solution?

There are very innovative and green emerging trends in the space of bioplastics. However, there remains a significant challenge in verifying the feasibility of these in terms of both scale and cost efficiency. For example, replacing 3% of plastic with plant-based plastic would consume 5% of the global corn crop. To replace current plastic demand by bioplastics would need almost the entire global production of corn and wheat, rendering them unavailable for food consumption. We need to focus on non-food renewable resources that do not compete with land and resources required for food production.

Figure 3.3 shows the production and capacity needed to fulfill plastics demand with food crops. The production of bioplastic is very small and scale-up challenges remain to be resolved from both a technological and a commercial viewpoint.

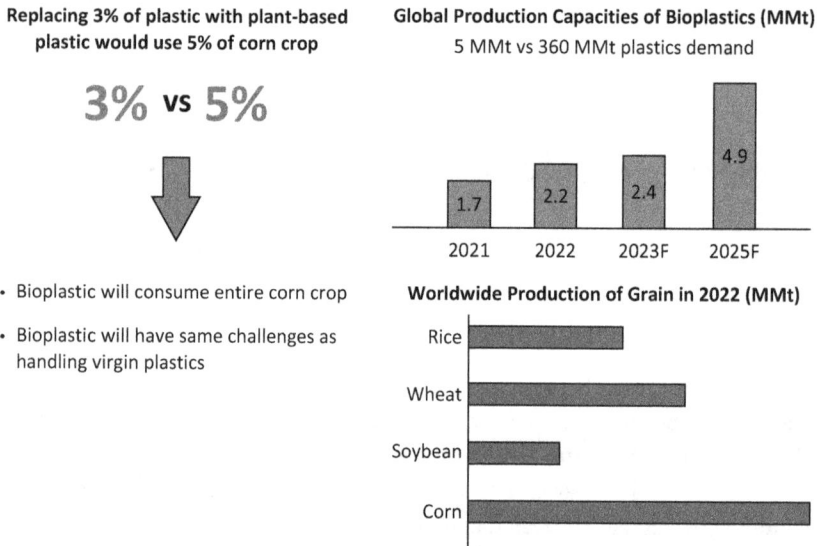

Figure 3.3 Production and capacity needed to meet plastics demand with food crops [5]

Bioplastics also do not fully solve the problem as they have the same challenges, such as recycling in the case of virgin plastics. In the case of biomaterials, biodegradability is a welcome feature but, in some cases, it impairs performance and durability. There are several innovative developments that focus on bio-based materials as valuable raw materials for sustainable textiles.

These innovations are promising, but will be slow to evolve and scale up. The urgent need to address the sustainability challenge requires a 72-fold increase in bioplastic capacity to meet current demand, without competing with food crops.

3.1.4 Solutions Need to be Double-Stitched

Besides the complexity of the textile supply chain, there are challenges such as the toxicity of dyes, the generation of microfibers and the end-of-life implications of synthetic materials.

Fortunately, there are numerous initiatives started by the textile leading brands, start-ups, universities, and research institutions who are trying to solve the problems facing our industry. We must double down on a far more collaborative and comprehensive approach to make sustainable textiles.

Over the last few years, it has been encouraging to see textile industry initiatives towards circularity and sustainability. Leading textile brands have made aggressive commitments to embrace circularity. Sustainability and circularity will define the design of garments that reduce demand for non-renewable resources, will cut down on waste, emissions, water pollution and the generation of microfibers. Listed in Figure 3.4 are different types of bio-fibers and bio-fabrics that are the result of work done by industry and research institutions.

1. PLA Fiber (polylactic acid fiber)
Obtained from polymerization of lactic acid.

2. Chitosan fiber
A natural biopolymer that is derived from chitin.

3. Algae-Based fiber
Algal fiber is completely eco-friendly and bio-degradable.

4. Bacterial cellulose fiber
Produced from cellulose which is extracted from bacterial cell walls.

5. Collagen fiber
A protein fiber, obtained from renewable natural sources.

6. Fiber made from spider silk
A protein fiber, which is being extracted from the glands of a spider.

7. Milk-weed fiber
Fiber obtained from a perennial plant of genus Asclepias.

8. Bio-wash/Enzyme wash fabric
An attempt to innovate a suitable textile processing method that delivers eco-friendly products.

9. Milk
Derived from casein protein, which is present in milk, also so it is named as milk or casein fiber.

10. Bacteria-Based fabric
Produced by microbes converting liquid bio-mass waste products from coconuts, beer, sugar and liquid food.

Challenges on technologies scale up, cost and quality

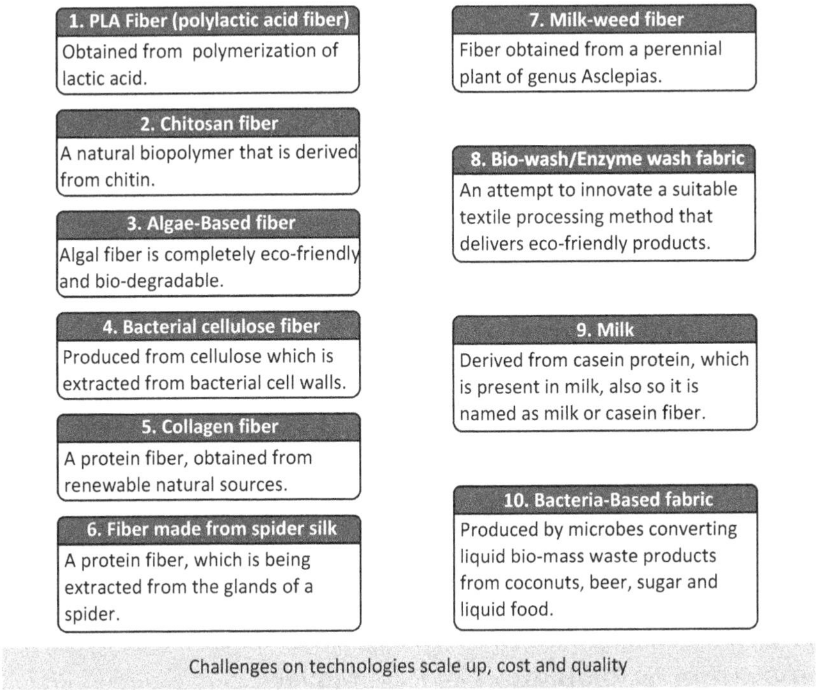

Figure 3.4 Different types of bio-fibers and bio-fabrics [6]

One such initiative that deserves to be mentioned is Biotexfuture, a European initiative to fund sustainable textiles projects. Their vision is to convert the textile value chain from petroleum-based to bio-based, however, there are challenges on technologies, scale up, cost and quality. The Biotexfuture projects are presented in Figure 3.5.

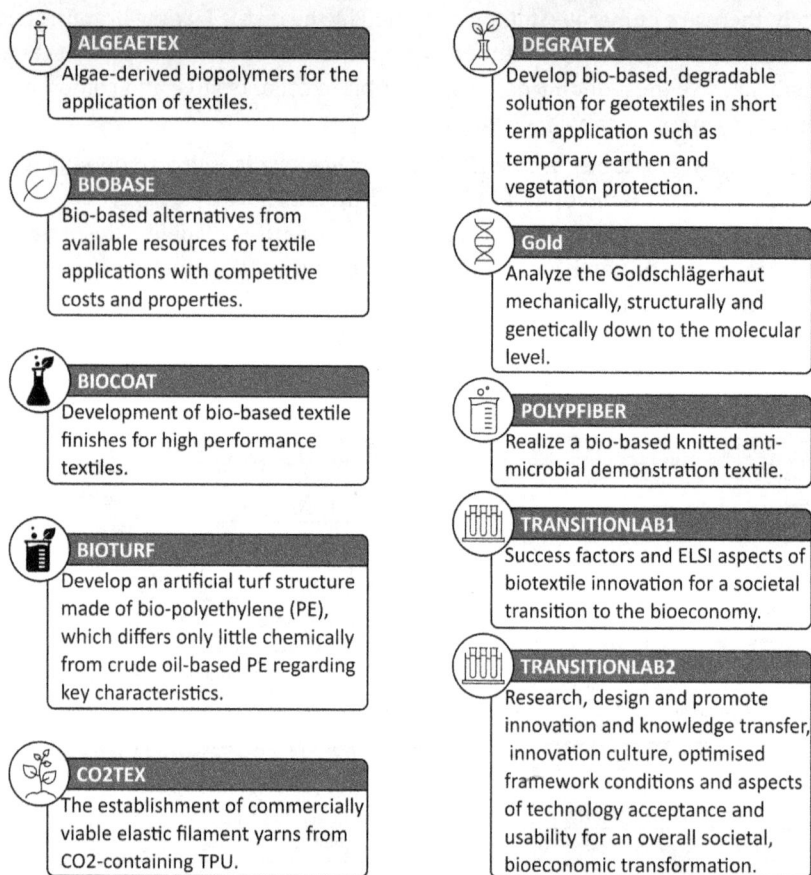

ALGEAETEX
Algae-derived biopolymers for the application of textiles.

BIOBASE
Bio-based alternatives from available resources for textile applications with competitive costs and properties.

BIOCOAT
Development of bio-based textile finishes for high performance textiles.

BIOTURF
Develop an artificial turf structure made of bio-polyethylene (PE), which differs only little chemically from crude oil-based PE regarding key characteristics.

CO2TEX
The establishment of commercially viable elastic filament yarns from CO2-containing TPU.

DEGRATEX
Develop bio-based, degradable solution for geotextiles in short term application such as temporary earthen and vegetation protection.

Gold
Analyze the Goldschlägerhaut mechanically, structurally and genetically down to the molecular level.

POLYPFIBER
Realize a bio-based knitted anti-microbial demonstration textile.

TRANSITIONLAB1
Success factors and ELSI aspects of biotextile innovation for a societal transition to the bioeconomy.

TRANSITIONLAB2
Research, design and promote innovation and knowledge transfer, innovation culture, optimised framework conditions and aspects of technology acceptance and usability for an overall societal, bioeconomic transformation.

Vision to convert the textile value chain from petroleum-based to bio-based

Figure 3.5 Overview of Biotexfuture projects [7]

3.1.5 Innovation or Regulation as First Mover

Policy plays a key role in this transformation and national governments have a vested interest in reversing this phenomenon.

Governments are increasingly conscious of the environmental impact of industrialization and resource use; consequently they have started introducing regulatory and policy framework to regulate industries more strictly, such as single-use plastics and textiles. In the private sector, some of the top textile brands have made ambitious commitments to use renewable and recycled materials ranging from 30 to 100%. Still, consumers remain unconvinced and expect governments and industry to intensify and accelerate efforts to protect the environment. This oscillation between regulation and innovation will continue to push nature positive solutions.

To help guide the way and create a common vision of this compelling need, the European Commission has initiated the Strategy for Sustainable and Circular Textiles.

"By 2030 textile products placed on the EU market are long-lived and recyclable largely made of recycled fibers, free of hazardous substances and produced in respect of social rights and the environment. Consumers benefit longer from high quality affordable textiles, fast fashion is out of fashion, and economically profitable reuse and repair services are widely available. In a competitive, resilient, and innovative textiles sector, producers take responsibility for their products along the value chain, including when they become waste. The circular textiles ecosystem is thriving, driven by sufficient capacities for innovative fiber-to-fiber recycling, while the incineration and landfilling of textiles is reduced to the minimum." Consumers expect governments and industry to intensify and accelerate efforts to protect the environment. Figure 3.6 shows a small selection of various policy initiatives on regulation by governments.

Waste2Fresh is an EU Horizon 2020-funded project aiming to develop and demonstrate a closed-loop recycling system for the dye wastewater from textile factories. France is the first country to adopt a law that requires all new washing machines to have microfiber filters by 2025.

China has enacted environmental policies to crackdown on textile dye pollution. Indeed, the policy landscape is changing, and the textile industry must adapt.

California, for its part, mandated recycled content targets of all beverage containers placed on the market of at least 25% by 2025, and 50% by 2030. The Golden State has adopted a first-in-nation approach for protecting ocean and human health from microplastics pollution.

Realizing this ambitious vision will require policy, investment, innovation and massive collaboration across the textile value chain and beyond.

While it is important to have regulations put in place, the next step is to solve the challenge of its execution when the bulk of the textile materials are imported and infrastructure for a circular economy near consumption centers is not ready. We need to encourage innovation, help create capacity, deploy technology for scalable solutions and undertake collective efforts to promote circularity.

EU	
	• Initiated strategy for sustainable and circular textiles on 30th of March 2022
	• Policy targeting single use plastics to collect 77% of beverage bottles by 2025 and 90% by 2029
	• Imposed a plastics tax of 800 euro/MT for increasing recycled content in packaging industry
	• ReHubs initiative to set up an integrated system based on 5 recycling hubs

California	
	• Recently passed Senate Bill, requires a 25% reduction of single-use plastic packaging, all packaging is to be recyclable or compostable and a 65% plastic recycling rate - all of which are due by 2032
	• Adopts a first-in-nation approach for protecting ocean and human health from microplastics pollution
	• Incentives and investments to businesses offering sustainable fashion services along with reduction of regulation and taxation for thrift stores

Netherlands	
	• The Green Deal Circular Procurement involves €100 million to pilot circular procurement, facilitating implementation and the removal of obstacles in regulation
	• A tax increase on natural gas and plans to price carbon emissions in combination with reduction of income tax and employers' costs (Sweden, Germany and Austria also)
	• Low VAT for circular products and services influences the choice of consumers and businesses for circularity

China	
	• Enacted environmental policies to crackdown on textile dye pollution
	• Beijing to promote low-carbon textile production, encourage using sustainable fibers, strengthen producers' social responsibility, improving the recycling system and raise investment in R&D for textile technology

Encourage innovation, create capacity, deploy technology
for scalable solutions to promote circularity

Figure 3.6 Overview of various policy initiatives to increase sustainability

3.1.6 Innovation Recognized across the World, but Most of them are Stand-Alone Initiatives

Fortunately, there are numerous initiatives started by the textile leading brands, start-ups, universities, and research institutions who are trying to solve the problems facing our industry. Alternative materials are key to reaching the goal of using recycled or sustainably sourced materials.

Some of the work being done by industry and institutes around cellulosic materials is shown in Figure 3.7. Alternative materials are key to reaching the goal of using recycled or sustainably sourced materials.

Infinited Fiber Company's Infinna™	The Hurd CO
- A cellulose carbamate fiber currently created out of 100% post-consumer textile waste.	- Engineers man-made cellulosic fiber pulp from 100% agricultural waste.

Asahi Kasei's Bemberg™
- A cupro fiber made in japan from 100% cotton linter, a pre-consumer residue from cotton processing.

Inspidere's Mestic ®
- A method in development to retrieve and convert cellulose from dairy cow manure into regenerated cellulose fibers.

Aalto University's Ioncell
- A technology in development that turns used textiles, pulp or old newspaper into new textile fibers, commercial production planed for 2025.

Birla's Liva Reviva
- A new viscose fiber made with up to 20% pre-consumer waste.

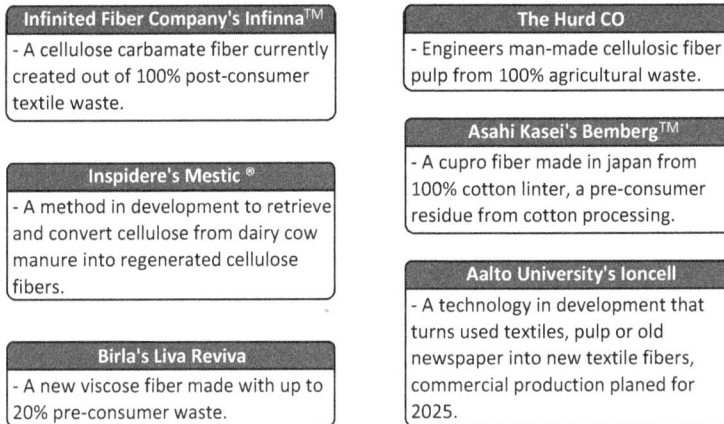

Commitments to the Changing Markets Road-map have been made, for example, by the members ASOS, C&A, Esprit, H&M, Inditex, Levi's, M&S, New Look, Next, Reformation and Tesco

Figure 3.7 Overview of innovations in the cellulosic material sector [6]

The stand-alone efforts employed by brands, industry, research institutions and individual governments will not be effective, as nature does not respect political boundaries. Therefore, a united global effort is needed to meet this challenge.

3.1.7 Chemical Fibers are Most Critical for Gaining Critical Mass to Sustainable Textiles

The market share of chemically recycled polyester is expected to grow in the coming years. For example, Adidas aims to replace all virgin polyester with recycled polyester in all Adidas and Reebok products, where a solution exists, by 2024. H&M has a target of only using recycled or other sustainably sourced materials by 2030, while IKEA is committed to ending the dependency on virgin fossil materials and using only renewable or recycled materials by 2030.

We must double down on a far more collaborative and comprehensive approach to make sustainable textiles. This should be driven by behavioral change and political action to create a culture of recycling and reuse.

Some of the developments and commitments by the polyester fibers industry are detailed in Figure 3.8. The market share of chemically recycled polyester is expected to grow in the coming years.

Some of the prominent industry players have committed to changing market roadmaps by pursuing alternative key materials to achieve the goal of recycled or sustainably sourced materials.

Recycled Polyester	Bio-based Polyester
Eastman Announced a new $250 M molecular recycling facility using textile feedstock in the USA. The new facility will use over 100 kt of plastic waste that can not be recycled by current mechanical methods. **Carbios** Produced the first clear plastic bottles from enzymatically recycled textile waste. Its first industrial unit with production capacity of 40 kt/y of recycled PET, expected to generate first revenues in 2025. **INDORAMA VENTURES & loop INDUSTRIES** JV has developed a patented chemical recycling process to depolymerize all kinds of polyesters with zero energy use. The chemical recycling produces recycled polyester DMT and MEG.	**FAR EASTERN GROUP** TopGreen® Bio PET Filament is bPET filament made with 30 % bio-based feedstock from sugarcane. **INVISTA** Launched LYCRA® T400® EcoMade fiber. >65% of the overall fiber content comes from a combination of chemically recycled PET bottles and renewable plant-based resources (corn). **INDORAMA VENTURES** Bio-PET resin made with 30 percent plant-base bio-MEG. **VIRENT** Completed a year-long run of demonstration plant which demonstrates the technology to convert plant sugars to bio-paraxylene, a critical raw material for bio-polyester fiber.

Adidas aims to replace all virgin polyester with recycled polyester in all Adidas and Reebok products, where a solution exists, by 2024.
H&M has the target to only use recycled or other sustainably sourced material by 2030.
IKEA is committed to ending the dependency on virgin fossil materials and using only renewable or recycled materials by 2030.

Figure 3.8 Developments and commitments in the polyester fibers industry [6]

Mechanical recycling of polyester has been a success, but cannot remove the color or separate contaminants from other materials. Chemical recycling technologies are developing fast, but technology, scale and cost remain a challenge. The industry and brands' engagement to develop and scale up chemical recycling is promising, but greater collaboration is needed to speed up such solutions.

For example, recycling polyamide helps decrease dependency on fossil-based raw materials and waste material. Organizations like VF Corporation have committed to increasing their uptake of recycled nylon to 50% by 2025. Details are given in the graphic below (Figure 3.9).

The efforts aimed at developing bio-based and recycled products that are now visible in specialized applications, such as elastane, polyurethane, polybutylene succinate, polylactic acid, and other bio-based polymers, serve certain niche segments, but have huge promise. Details are shown in (Figure 3.10).

Industry is engaged in low-carbon or carbon-neutral materials that will reduce the carbon footprint of the industry, but most of the work is still based on individual efforts. Once the industry collaborates as one unit, we can achieve a multiplier effect of industry capability and capacity. Other manmade fibers and materials are presented in Figure 3.11.

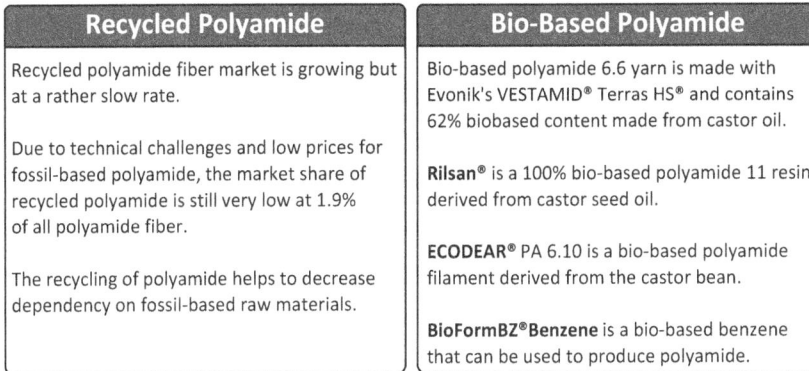

Recycled Polyamide	Bio-Based Polyamide
Recycled polyamide fiber market is growing but at a rather slow rate. Due to technical challenges and low prices for fossil-based polyamide, the market share of recycled polyamide is still very low at 1.9% of all polyamide fiber. The recycling of polyamide helps to decrease dependency on fossil-based raw materials.	Bio-based polyamide 6.6 yarn is made with Evonik's VESTAMID® Terras HS® and contains 62% biobased content made from castor oil. Rilsan® is a 100% bio-based polyamide 11 resin derived from castor seed oil. ECODEAR® PA 6.10 is a bio-based polyamide filament derived from the castor bean. BioFormBZ®Benzene is a bio-based benzene that can be used to produce polyamide.

Everlane is committed to using only recycled nylon by 2021.
VF Corporation has committed to increasing uptake of recycled nylon to 50% by 2025.
Volcom is committed to increasing their share of recycled nylon to 20% by 2020.

Figure 3.9 Overview of recycled and biobased polyamide [6]

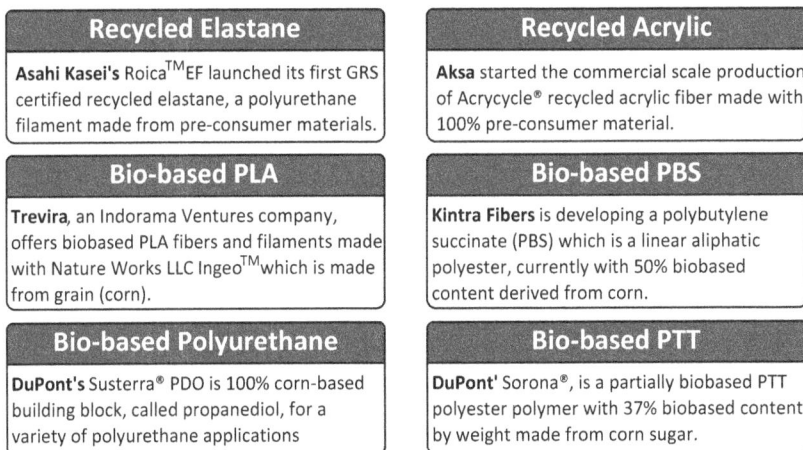

Recycled Elastane	Recycled Acrylic
Asahi Kasei's Roica^{TM}EF launched its first GRS certified recycled elastane, a polyurethane filament made from pre-consumer materials.	Aksa started the commercial scale production of Acrycycle® recycled acrylic fiber made with 100% pre-consumer material.

Bio-based PLA	Bio-based PBS
Trevira, an Indorama Ventures company, offers biobased PLA fibers and filaments made with Nature Works LLC Ingeo^{TM}which is made from grain (corn).	Kintra Fibers is developing a polybutylene succinate (PBS) which is a linear aliphatic polyester, currently with 50% biobased content derived from corn.

Bio-based Polyurethane	Bio-based PTT
DuPont's Susterra® PDO is 100% corn-based building block, called propanediol, for a variety of polyurethane applications	DuPont' Sorona®, is a partially biobased PTT polyester polymer with 37% biobased content by weight made from corn sugar.

Figure 3.10 Overview of recycled & bio-based developments in the chemical fibers industry [6]

We must align and address the challenges we are facing as one global effort and collaborate as one unit; otherwise, we will be wasting critical resources and time duplicating our efforts and compromising the power of the industry.

Recycled Fibers from Blended Textiles	Man-made Protein Fibers	CO_2-Based Fibers	Recycled and Bio-Based Non-Animal-Based Alternatives to Leather
Chemical recycling of blended materials	Bio-based manmade protein fibers are another material innovation but further research is required	Few companies are exploring innovative approaches to directly capture CO_2 from the air and use its carbon as feedstock for textiles.	While leather is a by-product of the meat and dairy industry, some brands prefer non-animal-based alternatives to leather.
Developed a technology to separate post-consumer polyester cotton blends on a molecular level and turn it into high-quality polyester pellets.	Bio-based protein fiber produced in a continuous spinning process which bio-based man-made "spidersilk" made of sugar, water, salts and yeast.	Covestro and its partners, have succeeded in making elastic textile fibers based on CO_2 and so partly replacing crude oil as a raw material.	Apple Peel Skin - A material which integrates apple peels into the skin of the material and reduces the amount of polyurethane in the material.
Developing a polycotton blend recycling process to separate polyester fiber from cotton, polyester monomers and dissolving pulp	Protein fiber made by genetically engineered silkworms and composed entirely of protein produced naturally by the silkworm.	Developing a carbon recycling technology, aim to create clothing like yoga pants from the CO_2 emissions from a steel mill.	Fruitleather Rotterdam - Developing a new process that converts left-over fruits into durable material, possibly strong enough to be used for shoes, handbags.

Figure 3.11 Manmade fibers and materials to increase sustainability of textiles [6]

3.1.8 Need for a Global Overarching, High-Impact and Scalable Program

Several programs are ongoing at different levels and are having an impact, but on a very small scale and it will take time before they are replicated in other regions. There is a need to have programs at a global level which can be rapidly scaled.

Different elements will play a crucial role in the success of such programs (Figure 3.12). A common thread encompassing these elements will help in attaining the desired speed.

An integrated approach will help in developing appropriate solutions.

Coordination
To achieve the desired
scale and speed, we need
greater coordination.

Integration
Several fragmented
projects need to be
integrated in reversing
the mega problem.

Response
Sense of urgency to solve
the textile circularity should
be similar to the response
we saw in case of Covid.

Governance
Global cooperation for
compliance and
discipline.

Behaviour
Habit-forming cultural
change for circularity.

Technology
Automation and digital
will accelerate the pace of
innovation scale-up.

Top Talent
Need to deploy the best
and the brightest talent in
the industry, research
institutes and government.

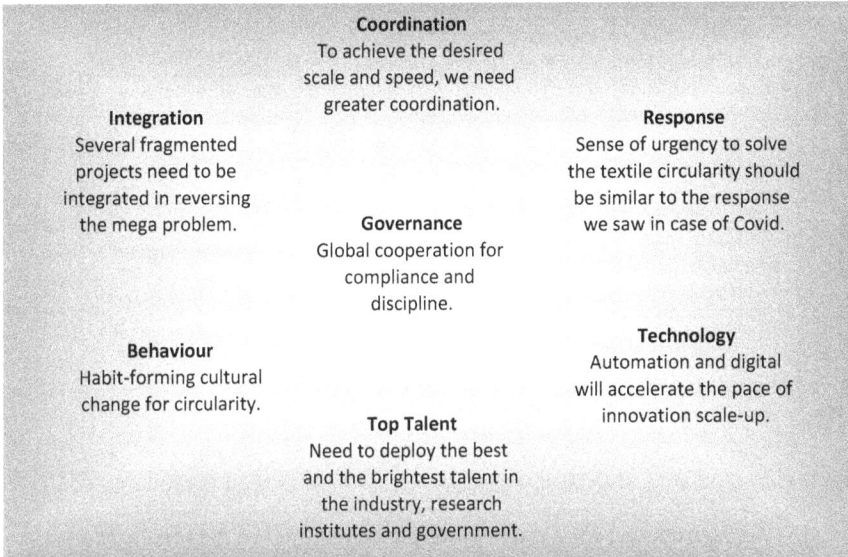

Figure 3.12 Overview of crucial elements for program success

3.1.9 Cultivating a New Mindset for Circularity

Over the last few years, it has been encouraging to see increased consumer awareness and textile industry initiatives aimed at sustainability. Leading textile brands have made aggressive commitments to embrace circularity. Consumer awareness and soon, purchasing behavior, will follow. Still, the bulk of these commitments are based on recycled bottle flakes and it is uncertain how far they can be sustained.

The problem facing us is so complex that it requires a new mindset for scalable, circular and environmentally friendly solutions (Figure 3.13). Sustainability and circularity will define the design of garments that reduce demand for non-renewable resources, will cut down on waste, emissions, water pollution and the generation of microfibers. Our industry must actively embrace new technologies that can deliver both performance and fashion without compromising on environmental impact, quality or pricing.

To transform the existing textile industry into a circular economy, we will have to address the complex supply chain that discourages circularity. In major consumption centers, such as North America and Europe, discarded garments end up in the landfill, as local industry do not yet have the matching capacity or skills to reprocess the textile waste.

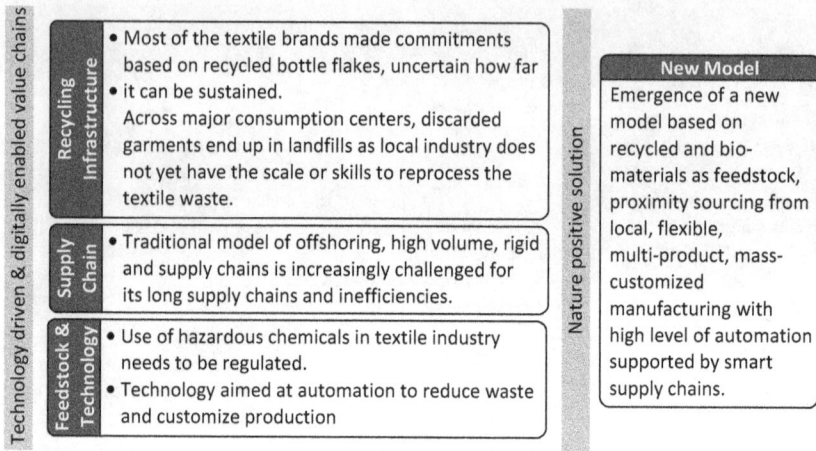

Figure 3.13 Mindset needed for a new nature-positive model of the textile value chain

We must move away from a traditional model of offshoring, high volume, rigid and made-to-stock goods which is increasingly challenged for its long supply chains and inefficiencies. Indeed, nearly 30% of the clothes shipped from Asia to North America are destroyed, because they never sell. We must transform the market by developing short supply chains backed by intelligent processes and smart factories. Additionally, ensuring prompt deliveries will help create a local supply chain eco-system.

Such macro drivers have led to the emergence of a new model which will place the emphasis on proximity sourcing from local, flexible, multiproduct, mass-customized manufacturing with a high level of automation supported by smart supply chains. Technology can speed up this process of localization in Western consuming centers by reducing wage gaps through automation and skills gaps through artificial intelligence and machine learning, both of which are efficient in terms of environmental and economic costs.

3.1.10 The Way Forward for Sustainable Textiles

Like SDG17 around Partnership, driving market transformation in the textile industry requires massive collaboration. Rising to the challenges described above, it is encouraging to see significant efforts by the industry to address the climate impact of textiles. One example, amongst many, is the Global Alliance for Circular Economy and Resource Efficiency (GACERE). To realize the necessary market transformation, we need to address the following challenges collaboratively and comprehensively:

- Encourage new circular products by providing necessary functions/incentives
- Lower taxes on circular materials to influence the choices made by consumers

- Create a platform among stakeholders in the circular space to share new ideas, practices, and upcoming regulations

- Consumer awareness – Promote sustainable products and communicate the benefits

- Traceability – Release standards and criteria for adherence to circularity content in appropriate response to climate change

- Encourage start-ups, universities, and organizations involved in innovation and R&D

- Emergence of new model for scalable, circular, and nature-positive solutions

- Promote automation and digital tools by incentivizing them to operate at scale

- Invest in innovation and technology to match circular products with virgin materials

To successfully transition from a resource-intensive textile model to a new low-impact value chain designed on circular principles, we need to be more innovative, nature positive, technology driven, resolute, socially fair and end-to-end collaborative. While significant efforts to protect the environment are being initiated across the industry, there is a need for a uniform global textile eco-code that brings discipline on execution and disclosures on compliance.

Indorama Ventures Limited is doing its part. Building on our decade-long practice of disclosure and sustainability reporting, we have recently made announcements about significant investment in mechanical recycling infrastructure of $1.5 billion as well as chemical recycling and the increasing use of bio-based feedstocks of $8 billion through 2030 (see Figure 3.14).

Indorama Venture Limited (IVL) commitments and goals

- Achieve 25% bio-based or circular feedstock by 2030

- Invest $8 billion in developing this pathway

- $1.5 billion committed to bottle-to-bottle recycling capacity of 1 million tons by 2030

Working with several industry partners to achieve circular economy for plastics and intending to play a leading role by bringing customers of recycled products into the value chain.

Figure 3.14 Commitments and goal of Indorama Ventures Limited aimed at increasing recycling capacity and building a bio-based or circular feedstock pathway

References for Section 3.1

[1]   Oil demand by sector and scenario to 2030; *https://www.iea.org/data-and-statistics/charts/oil-demand-by-sector-and-scenario-to-2030*, last accessed 11.06.2025

[2]   Fashion Waste Facts and Statistics; *https://www.businesswaste.co.uk/your-waste/textile-recycling/fashion-waste-facts-and-statistics/*, last accessed 11.06.2025

[3] The Future of Petrochemicals, Towards more sustainable plastics and fertilisers: International Energy Agency, 2018, *https://iea.blob.core.windows.net/assets/bee4ef3a-8876-4566-98cf-7a130c013805/ The_Future_of_Petrochemicals.pdf*, last accessed 11.06.2025

[4] Textile Exchange and Fashion Industry Charter for Climate Action launch the 2025 Recycled Polyester Challenge, a joint industry initiative: 2021 *https://textileexchange.org/news/textile-exchange-and-fashion-industry-charter-for-climate-action-launch-the-2025-recycled-polyester-challenge-a-joint-industry-initiative/*, last accessed 11.06.2025

[5] Alfred Rudin, Phillip Choi, Chapter 13 – Biopolymers, Editor(s): Alfred Rudin, Phillip Choi, The Elements of Polymer Science & Engineering (3rd Edition), Academic Press, 2013, Pages 521–535, ISBN 9780123821782, *https://doi.org/10.1016/B978-0-12-382178-2.00013-4.*

[6] Preferred Fiber & Materials, Market Report 202: Textile Exchange, 2021, *https://textileexchange. org/app/uploads/2021/08/Textile-Exchange_Preferred-Fiber-and-Materials-Market-Report_2021.pdf*, last accessed 11.06.2025

[7] BIOTEXFUTURE, Projects: *https://biotexfuture.info/projects/*, last accessed 11.06.2025

3.2 The Textile Waste Value Chain

Thomas Böschen, TEXAID Beteiligungsverwaltung Deutschland GmbH, Darmstadt

3.2.1 Five Forces Shape the Industry

The textile value chain encompasses the various stages from production and distribution of textiles, starting from the sourcing of raw materials for fiber production right up to the sale of a finished textile product. Each stage of the textile value chain increases the value of the raw materials, resulting in finished products ready for consumers to buy. The textile value chain is currently somewhat linear, as too little textile waste is fed back into the value chain. In the coming years, the European textile waste value chain will be shaped by five forces. To enable textile circularity and improve sustainability, significant efforts are needed, especially within the collection, sorting, preparation for recycling and recycling.

A Broader Look: The EU Strategy for Sustainable and Circular Textiles

In the Green Deal, unveiled in 2019, the EU adopted the goals of Agenda 2030 and the Paris Climate Agreement, presenting a comprehensive European strategy for achieving these objectives. The EU's Green Deal aims to foster a carbon-neutral and sustainable economy in Europe by 2050, thereby mitigating environmental impact. If this is to be achieved, it is essential to significantly curtail greenhouse gas emissions in Europe. The EU has set a target of a minimum 55% reduction in emissions by 2030, compared with 1990, signaling a markedly more ambitious objective than the previous EU target of a 40% emissions reduction by 2030.

In 2020, the Circular Economy Action Plan (CEAP) was introduced as part of the current version of the Green Deal. The CEAP aims to enhance resource efficiency and reduce waste by making products more durable and promoting reuse. This is achieved through a multitude of measures, including optimizing production processes, advocating for repair and reuse services, and incentivizing the use of recycled materials. An overview of the EU strategies for sustainable and circular textiles is given in Figure 3.15.

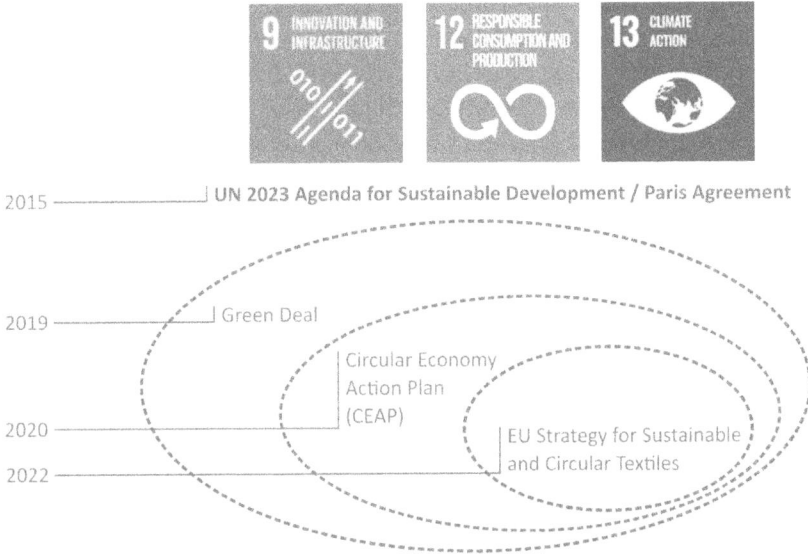

Figure 3.15 EU strategies for sustainable and circular textiles

As mentioned before, five major forces will shape the European textile waste value chain in the coming 5 to 10 years. These five forces are as follows:

8. EU Regulations Demand Higher Collection Quota for Textile Waste

The EU Strategy for Sustainable and Circular Textiles, published in 2022, represents an action plan by the European Commission that supports the implementation of the goals of the European Green Deal and aims to guide the textile industry toward a more sustainable future. The textile strategy features important and binding requirements.

Under the current EU rules on waste, member states are required to set up separate collection of textiles by January 1, 2025. This involves strengthening separate collection, sorting, reuse, and recycling capacities within the EU. The Commission has proposed introducing mandatory and harmonized Extended Producer Responsibility (EPR) schemes for textiles in all member states. These schemes require producers to take responsibility for the entire life cycle of their products, particularly at the end of the product's life. The proposal also aims to foster research and development in innovative technologies that promote circularity in the textile sector [1].

Regulatory Advances Limit the Export of Unsorted Textile Waste

The Waste Shipment Regulation aims to prevent the transfer of waste from developed to developing countries. The European Union has proposed an amendment to the Waste

Framework Directive that targets textile waste. The proposal aims to bring about a more circular and sustainable management of textile waste, in line with the vision of the EU Strategy for Sustainable and Circular Textiles. The proposal will foster research and development in innovative technologies that promote circularity in the textile sector. It also supports social enterprises involved in textile collection, sorting, reuse, and recycling, and will ultimately incentivize producers to design more circular products. To reduce illegal waste shipments to non-EU countries, often disguised as intended for reuse, the Commission's proposal further clarifies the definitions of waste and reusable textiles. This will complement the directive on waste shipments, which ensures that textile waste is only exported when there are guarantees that the waste is managed in an environmentally sound manner [2, 3, 4].

Breakthroughs in Sorting Technology

Initial advances have been made in textile waste sorting through the introduction of automation and artificial intelligence (AI). Innovative technologies currently in development seek to automate the classification, sorting, and digitization of textile waste sorting. In the area of sorting for recycling, the first fully automated sorting machines have been installed to automatically sort textiles by color and fiber-composition using near-infrared light. Köppe et al. 2023 found that innovative solutions for sorting textile waste are available on the market, although with different technology readiness levels (TRLs) and are therefore difficult to combine on a commercial scale today. The TRL of automatic sorting for recycling is much more advanced than sorting for reuse. Furthermore, there are still technical hurdles within sensor technologies for sorting for recycling to overcome [5, 6, 7].

Growth of Second-Hand Retail

The growth of the second-hand textile retail market has been significantly driven by environmentally aware consumers. This market has seen a dramatic increase in sales, with the global resale apparel market projected to grow to 51 billion U.S. dollars by 2026 [8]. Consumers' mounting environmental concerns are contributing to this growth, with 70% of preowned buyers liking the sustainable aspect of second-hand consumption. The trend toward second-hand clothing has the potential to achieve a market share of up to 20% over the next decade, thus becoming a significant market segment in fashion retail. The second-hand fashion market is growing at a much faster rate than sustainable fashion, with consumers turning to resale more and more [9, 10, 11, 12].

Emergence of Textile-to-Textile Recycling Technologies

The emergence of textile-to-textile recycling technologies marks a significant advancement in the textile industry. These technologies focus on fiber-to-fiber recycling, turning textile waste into new fibers that are then used to create new clothes or other textile products. Some technologies, like mechanical recycling of pure materials, are already established, while others, like chemical recycling of cotton to cellulose or polyester, are on the brink of commercialization. These innovative solutions improve circularity in the fashion industry [13, 14].

A graphical overview of the five forces is given in Figure 3.16.

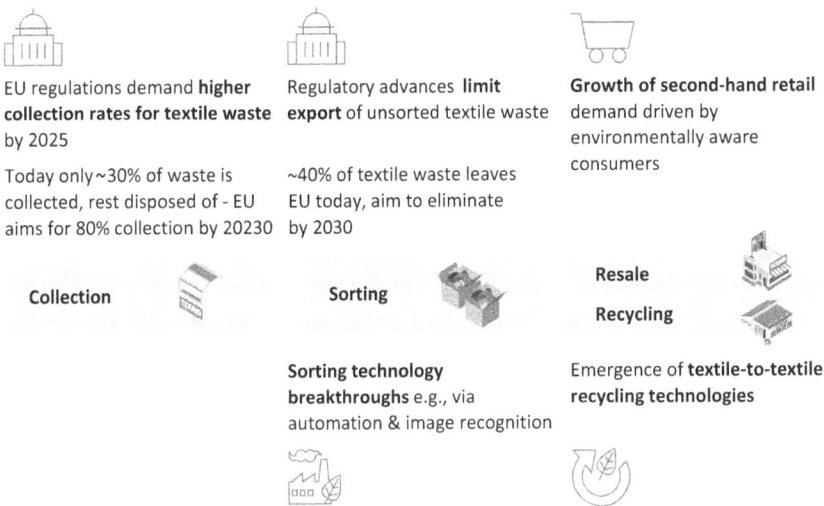

EU regulations demand **higher collection rates for textile waste** by 2025

Today only ~30% of waste is collected, rest disposed of - EU aims for 80% collection by 20230

Regulatory advances **limit export** of unsorted textile waste

~40% of textile waste leaves EU today, aim to eliminate by 2030

Growth of second-hand retail demand driven by environmentally aware consumers

Collection

Sorting

Resale

Recycling

Sorting technology **breakthroughs** e.g., via automation & image recognition

Emergence of **textile-to-textile recycling technologies**

Figure 3.16 The five forces according to [15]

3.2.2 Textile Waste Levels Continue to Increase, Putting Pressure on Collection and Sorting of Textile Waste

The initial forces induced by regulations are leading to a noteworthy increase in collection and sorting volumes. There are no consistent figures or statistics regarding the quantity of used textiles, or the current pathways for collection, sorting, and recycling. Consequently, the quantities can only be estimated in a broad sense.

Worldwide

In 2022, the global production volume of textile fibers was approximately 113.8 million metric tons. This included 25.2 million metric tons of natural fibers such as cotton or wool, and 87.6 million metric tons of chemical fibers, which include synthetic fibers such as polyesters or polyamides, and manmade cellulosic fibers like viscose or rayon [16].

The latest forecasts point to a significant increase in fiber production during the next years. Global fiber production is expected to reach 149 million metric tons by 2030. This translates into a production per capita of approximately 17.5 kilograms per person in 2030 in comparison to 14 kilograms per person a decade earlier [17].

It is estimated that, out of the 114 million tons of textile fibers, more than 100 million tons of textiles are manufactured. Over 60% of these textiles are attributed to the apparel and home textile sectors. Currently, the majority of textiles follow a linear model, being produced through a highly resource-intensive process and utilized for a short time by the end-user.

It is estimated that only around 20% of the 100 million tons of textiles are collected, sorted, and partially recycled. Approximately 80% of textiles are either directly incinerated or deposited in landfills.

Europe

In Europe, it is anticipated that the quantity of post-consumer household textile waste will increase from approximately 7 million tons to up to 9 million tons by 2030. In 2022, an estimated 2.3 million tons of post-consumer textiles were separately collected in Europe, accounting for approximately 33% of the total volume brought into the market. Due to the growth in quantities, particularly driven by fast fashion and the demand for separate collection of textiles outlined in the Textile Agenda, it is projected that the separately collected amounts will rise to up to 3.7 to 5.8 million tons by 2030.

Of the 2.3 million tons collected in Europe in 2022, around 0.9 million tons, or roughly 40%, are exported without sorting to non-European countries. Given the expected growth in quantities and without further investments in sorting and recycling facilities, the export share of unsorted textile waste is expected to exceed 50%.

This does not align with the vision of the EU as outlined above. It is estimated that by 2030 a 2.5-fold increase in European sorting capacity is necessary (Figure 3.17). Therefore, major investments in sorting facilities are required if future needs are to be met. Estimates range from 50 to 120 new sorting facilities are needed by 2030, an increase of up to 350%.

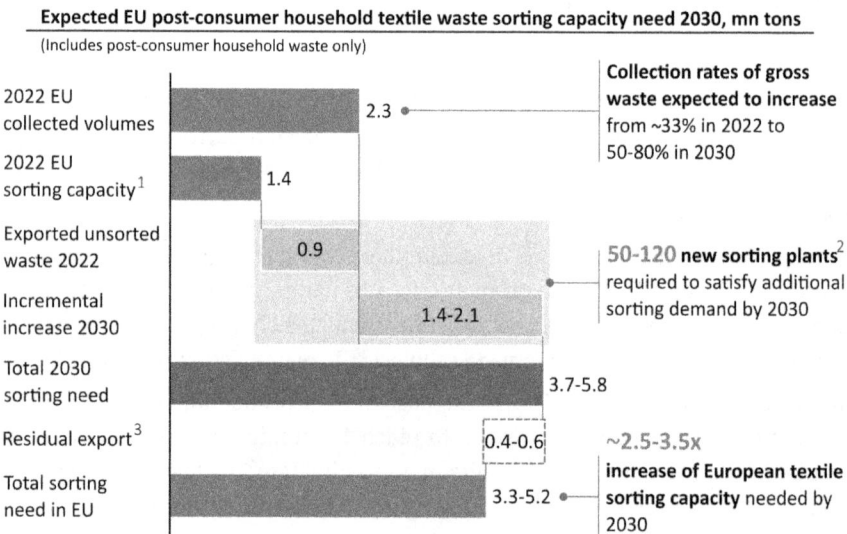

Expected EU post-consumer household textile waste sorting capacity need 2030, mn tons
(Includes post-consumer household waste only)

2022 EU collected volumes	2.3 — Collection rates of gross waste expected to increase from ~33% in 2022 to 50-80% in 2030
2022 EU sorting capacity [1]	1.4
Exported unsorted waste 2022	0.9 — 50-120 new sorting plants [2] required to satisfy additional sorting demand by 2030
Incremental increase 2030	1.4-2.1
Total 2030 sorting need	3.7-5.8
Residual export [3]	0.4-0.6 — ~2.5-3.5x increase of European textile sorting capacity needed by 2030
Total sorting need in EU	3.3-5.2

1: Assuming plants of 25-50kt sorting capacity
2: EU sorting capacity based on assumption that currently 40% of unsorted textile waste is exported
3: Assuming 10% leakage of theoretical sorting need

Figure 3.17 Expected EU need for sorting capacity for post-consumer household textile waste by 2023 according to [15]

3.2.3 Recycling Value Chain Starts with Collecting and Sorting Textile Waste

To collect and sort the increasing volume of textile waste and to prepare the textiles for an efficient resale or a high value recycling, a well-structured and sophisticated textile recycling value chain is essential. Such a value chain needs professional collection, sorting and preparation. A typical resale and recycling value chain is shown in Figure 3.18.

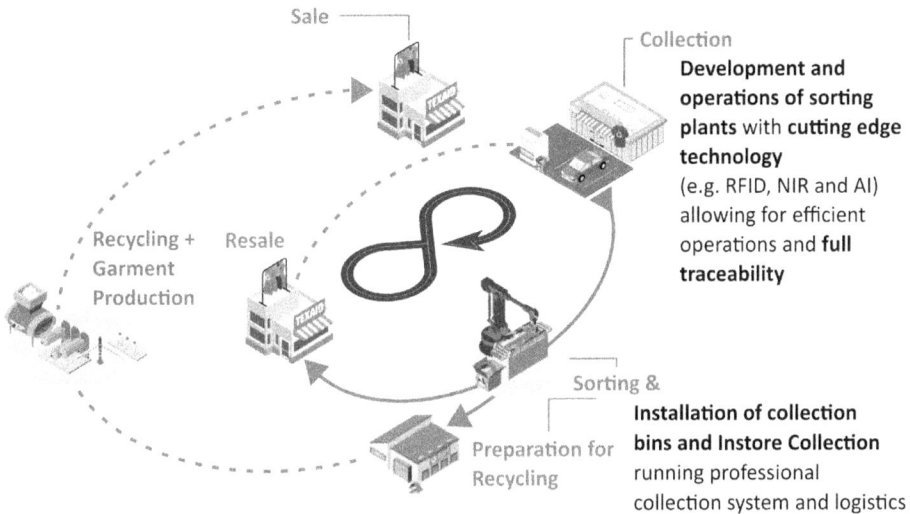

Figure 3.18 The resale and recycling textile value chain

The aim of textile recycling is to collect the highest-possible proportion of textiles and prepare them for reuse in different loops, ensuring optimal utilization based on their current state and condition.

In most countries of the world, there are regulations regarding the collection of textile waste. This is also true for the textile waste collection in the European community. However, there is at this stage no general requirement that is applicable to every country; instead, each country has different requirements to fulfill and permits that need to be applied for. In Germany, for example, companies have to follow different regulations and laws. The most important are:

- Waste Framework Directive (Abfallrahmenrichtlinie 2008/98/EG)

  This defines the main waste-related terms and contains key provisions for German waste disposal law. It also contains the EU waste hierarchy (Figure 3.19) which has to be followed. It serves as the foundation of EU waste management and introduces an order of preference for the managing and disposing of waste. The directive also introduces principles such as the "polluter pays principle" and "extended producer responsibility." It sets targets for EU countries to increase the reuse and recycling of waste materials. The waste hierarchy aims to minimize the adverse impact of waste generation and management.

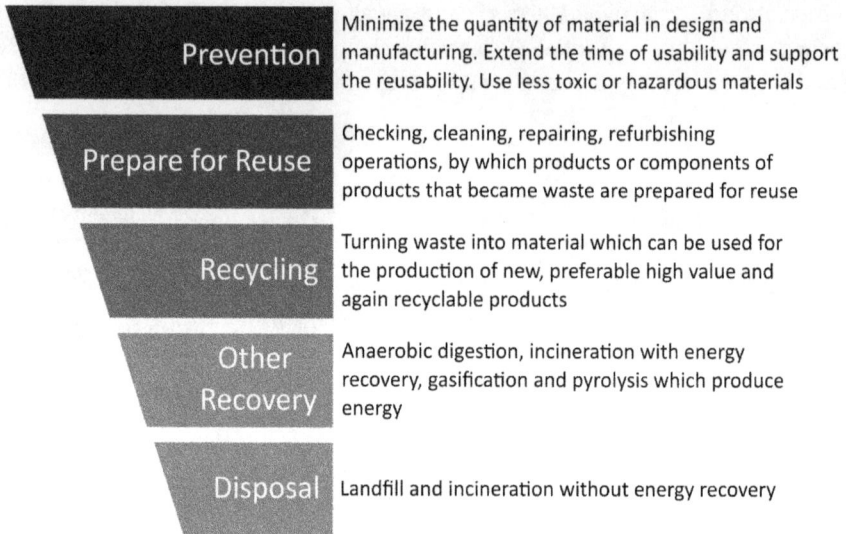

Prevention	Minimize the quantity of material in design and manufacturing. Extend the time of usability and support the reusability. Use less toxic or hazardous materials
Prepare for Reuse	Checking, cleaning, repairing, refurbishing operations, by which products or components of products that became waste are prepared for reuse
Recycling	Turning waste into material which can be used for the production of new, preferable high value and again recyclable products
Other Recovery	Anaerobic digestion, incineration with energy recovery, gasification and pyrolysis which produce energy
Disposal	Landfill and incineration without energy recovery

Figure 3.19 The EU waste hierarchy

- Circular Economy Act (Kreislaufwirtschaftsgesetz, KrWG)

  This is currently Germany's most important waste disposal law and the successor to the KrW-/AbfG law. According to the KrWG, any collection of waste must be reported to the relevant authority of each administrative district or independent city, of which there are more than 400 in Germany. The notification needs to be submitted at least three months in advance of the collection and must also include detailed information about the recycling of textile waste.

- State laws of the federal states and the municipal waste disposal law

  The Circular Economy Act (KrWG) is further differentiated by the waste management laws of the federal states. The collection and recycling of household waste at the municipal level is regulated by municipal ordinances.

In addition, any company aiming to collect textiles is asked to be certified as a waste management company, which comes along with a yearly review of the managed waste, recycling routes, checks on the reliability of the management etc.

3.2.3.1 Collection

There are several methods for collecting textile waste, each with a unique approach and benefits:

Curbside Collection

This method is similar to the collection of other recyclables like paper and plastic. Some municipalities offer this service, in which residents can leave their textile waste at the curb for pickup on designated days. It is often conducted once or twice a year by charity organizations using volunteers. The importance of the curbside collection is declining, especially as other collection methods, such as collection bins, are put in place.

Quality-wise, curbside collection delivers higher-quality textile waste since it is well packed by the consumer, mainly in garbage bags, and collected without delay by the collection company; this avoids cross-contamination. However, a lot of personnel is needed for curbside collection and so it also comes with high collection costs.

Drop-Off or Collection Bins

Collection by means of drop-off containers or collection bins is the most common way of collecting textile waste in most countries. The bins are often placed on public ground next to other collection boxes for paper or glass or on private ground often frequented by the public, such as shopping centers or community centers. People can drop off their unwanted textiles at their convenience. These bins are typically managed by public or private waste management companies or charitable organizations.

The quality which can be generated using collection bins is usually lower than of the other collection methods. This is mainly because the anonymity of the disposer reduces the care taken with regard to the textiles. Also, collection bins are often misused to dispose of non-textile waste materials, which leads to cross-contamination of the textiles.

Instore and Online Retail Take-Back Programs

In 2013, H&M was the first company to install a retail take-back program. Today, several fashion retailers, such as Zara, C&A, Primark, and Decathlon, offer programs where customers can return their old clothes, in some cases even in exchange for discounts on new items. These programs therefore not only encourage recycling, but also promote customer loyalty.

Online collection complements the take back programs of physical retailers as well as online retailers. The customer has the possibility to send in their used textiles using the return logistics of parcel services.

The quality of the instore and online take-back programs is also higher than the quality of the collection bins for the reasons mentioned earlier. The cost of collection in this channel is heavily dependent on the logistics structure of the retailer in the case of instore collection, and on the parcel prices in the case of online collection. In any event, the current costs of these collection channels are typically higher than those of the other collection methods.

Charity Shops, Donation Centers and Collection Events

These establishments accept donations of used clothing, which are then resold, reused, or recycled. This method not only diverts waste from landfills, but also supports charitable causes. Some communities or organizations hold special events where people can bring their unwanted textiles. These events are often held in conjunction with other recycling or waste collection initiatives.

Since it is a direct donation of textiles, the quality of this collection method is the highest.

The following pie chart (Figure 3.20) shows the percentage share of the above-mentioned collection methods as proportion of the total collection streams.

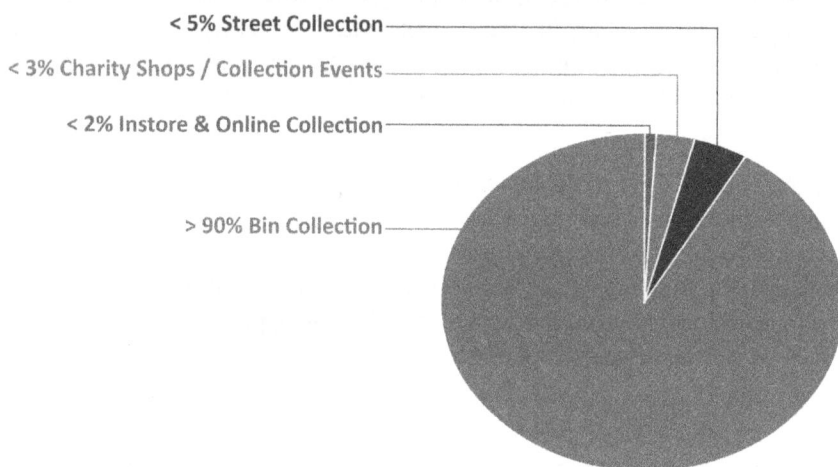

< 5% Street Collection

< 3% Charity Shops / Collection Events

< 2% Instore & Online Collection

> 90% Bin Collection

Figure 3.20 Collection streams

Logistics

The collection channel determines level of care and sustainability required for the collection and transport. This is crucial, especially for bin collection, which provides the highest share of the collection volume. Currently, the collection of textile waste from collection bins is done manually. Employees of the waste management company open the collection bin and empty the textile waste by hand, repacking loose material and separating impurities and any non-textile material to prevent cross-contamination.

To avoid unnecessary GHG emissions, transports of the textile waste should be kept to a minimum. In other words, sorting should be conducted in a nearby professional sorting facility. Also, it is crucial to optimize the logistics through the use of IT to optimize the emptying intervals and the driving routes between the collection bins.

3.2.3.2 Sorting

Sorting is the core element in the textile recycling chain; further processing of the collected goods without detailed sorting is not feasible. At the same time, sorting can be viewed as a reverse picking process. From a product mix, which may also contain a significant number of impurities, the sorting process aims to produce products that are pure in terms of their type and composition.

In a full sorting process, used textiles are sorted into hundreds of different products and marketed worldwide. This makes it possible to achieve reuse rates of around 60 percent, i.e., textiles can be marketed further in line with their original purpose through the second-hand market. About 30% of the textiles no longer meet the requirements of the second-hand market, due to their condition or to fashion, but can be recycled. About 10% of the sorted quantities is non-recyclable on account of contamination or material composition and so is thermally recycled.

To achieve these reuse and recycling rates, each textile is guided through a multi-stage decision chain. Owing to the complexity and inhomogeneity of the incoming goods and the high number of variables, it is not possible to automate most of the sorting steps today. As a result, sorting facilities are dependent on a large workforce of skilled sorters; increases in efficiency are currently only achieved in the upstream and downstream processes, such as automated intralogistics and packaging / warehousing.

The type and quality of the textiles determine the number of manual sorting steps, which can be up to four, in a full sorting process. Only through this complex procedure can it be ensured that the textiles intended for the second-hand market are adequately prepared for local sales and export in a manner suitable for the target countries, thereby avoiding the export of textile waste. Although such a sorting process guarantees high reuse rates and effective separation of the recycling components, and can thus be considered the most ecologically sensible sorting, it is usually only carried out by large sorting plants that have diversified sales channels for all fractions.

The structure of a typical sorting facility is as follows:

Incoming Goods Warehouse

Incoming goods are stored either in storage boxes or in the incoming transportation units, such as trailers or swap bodies. Since the collection quantities can fluctuate, due to weather conditions as well as other occasions, e.g. holidays, it is necessary to keep a stock level of several days of production.

Unloading

Ideally, the transportation units are unloaded automatically, saving on staffing levels and reducing the risk of accidents. Since the quality of textiles from different collection areas can vary, it is also important to mix the incoming goods during the unloading process. This ensures that textiles from, for example, stronger regions are combined with those from weaker regions, resulting in an average and stable quality which is then forwarded to the sorting process.

First and Second Presorting

This sorting step can be conducted at one sorting station or it can be separated into different sorting stations. In the latter case, the main task of first presorting is to open up the collection bags, prescreen the incoming goods and separate obvious waste, which may be household waste, but also moldy or other contaminated textiles. In addition, shoes and other items which usually are paired (such as suits) are tied together such that they stay paired. To an extent depending on the intralogistics, large items, such as blankets and curtains, are separated to avoid blocking the conveyor belts. At the second presorting station, the incoming goods are separated by product type. A quality check of the goods is not conducted at this stage. The size and structure of the sorting facility determine how the workforce at this stage has to divide the incoming goods, which may fall into more than 30 different categories (mostly consisting of different products or "product families").

Quality Sorting

After presorting, the product families are transported to the quality sorting stations. Skilled employees now have to decide whether the textiles are reusable or whether their condition fails to meet a market demand. In the latter case, they will be put into the recycling stream. The complexity of this sorting can be demonstrated by the example of the sorting process for a shirt (Figure 3.21).

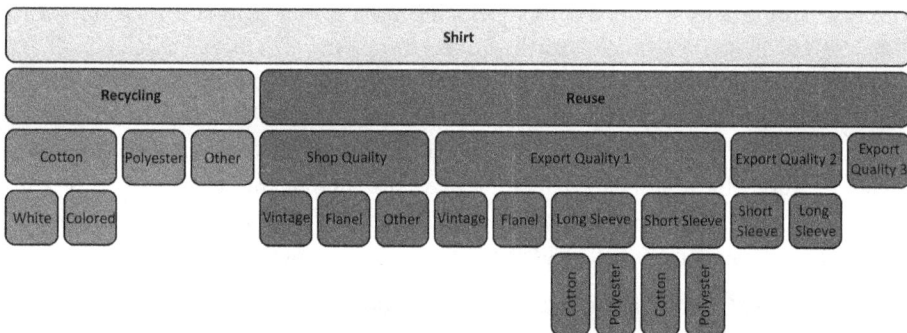

Figure 3.21 The decision tree for sorting a shirt

The quality the shirt determines whether it will be folded and sold on a "by piece" or a "by weight" basis. In general, once sorted by quality and category, the textile will be forwarded to the packaging, where further sorting and preparation for recycling or forwarding to the shop-sorting are performed.

Shop Sorting

Only the large sorting facilities conduct shop sorting. In this sorting step, the highest-quality textiles selected during the quality sorting stage are rechecked to ensure they are in good condition and functioning properly.

To ensure they remain in good condition, the garments are folded and placed in small boxes. This also aids in storage, as garments often arrive in the sorting facility out of synch with the current sales season and therefore have to be stored until the next season.

Preparation for Recycling

The field of recycling can currently be divided into downcycling and high-quality recycling. While downcycling does not necessarily require precise distinction of the fiber composition, textiles have to be sorted and further processed according to fiber type and other criteria to enable high-quality recycling.

To implement high-quality recycling in the sense of fiber-to-fiber recycling, very precise sorting of the textile starting material is required, along with conscientious removal of foreign materials such as buttons and zippers. The current state of the art is sorting with the aid of near-infrared spectroscopy (NIR). This technology uses the spectral signature of materials. Near-infrared light is radiated into the textiles and the reflected light spectrum is analyzed and compared with a database of known spectral signatures to identify and sort the material.

Packaging and Distribution

Packaging depends on the product type and destination. Large bales (up to 500 kg) are used for recycling material and waste. Export-quality textiles are usually packed into small bales (45–55 kg) for sea-container transport. Bags (45–55 kg) are used for truck transport. As already mentioned above, shop-quality textiles are usually stored within cardboard boxes.

A complete overview of the sorting process in a sorting facility is shown in (Figure 3.22).

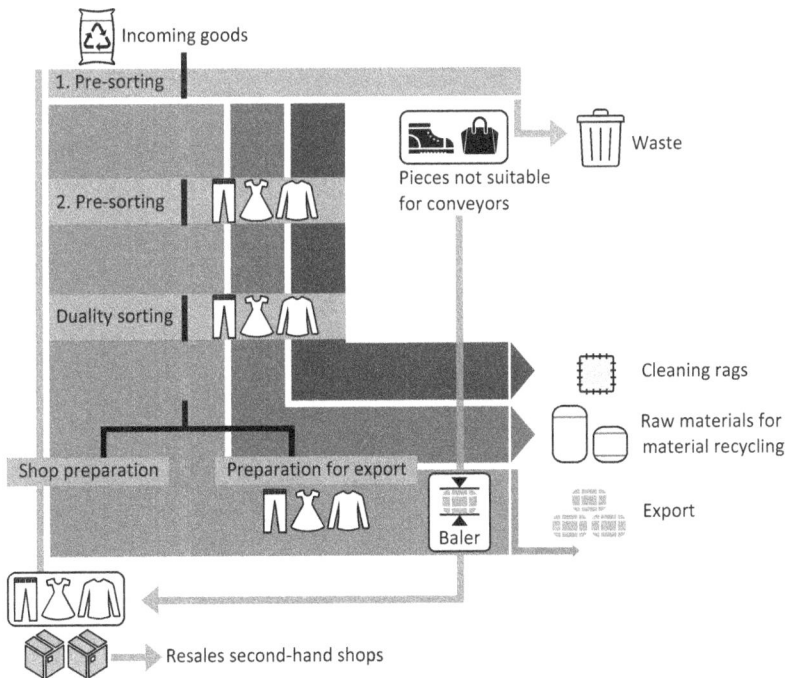

Figure 3.22 Process overview of a sorting facility

3.2.3.3 Projects to Support the EU's Vision of a Sustainable Textile Value Chain

There are numerous projects focused on realizing the vision of the textile strategy. One of the biggest challenges in the coming years is establishing the necessary infrastructure. The ReHubs initiative, initiated by the European Apparel and Textile Confederation "Euratex", which envisages fiber-to-fiber recycling of 2.5 million tons by 2030, has identified gaps particularly in collection and sorting as well as the recycling industry. Without high-quality collection of textiles and precise sorting, fiber-to-fiber recycling is not possible. Furthermore, capacities currently available in Europe are far from sufficient to supply a significant recycling industry with material.

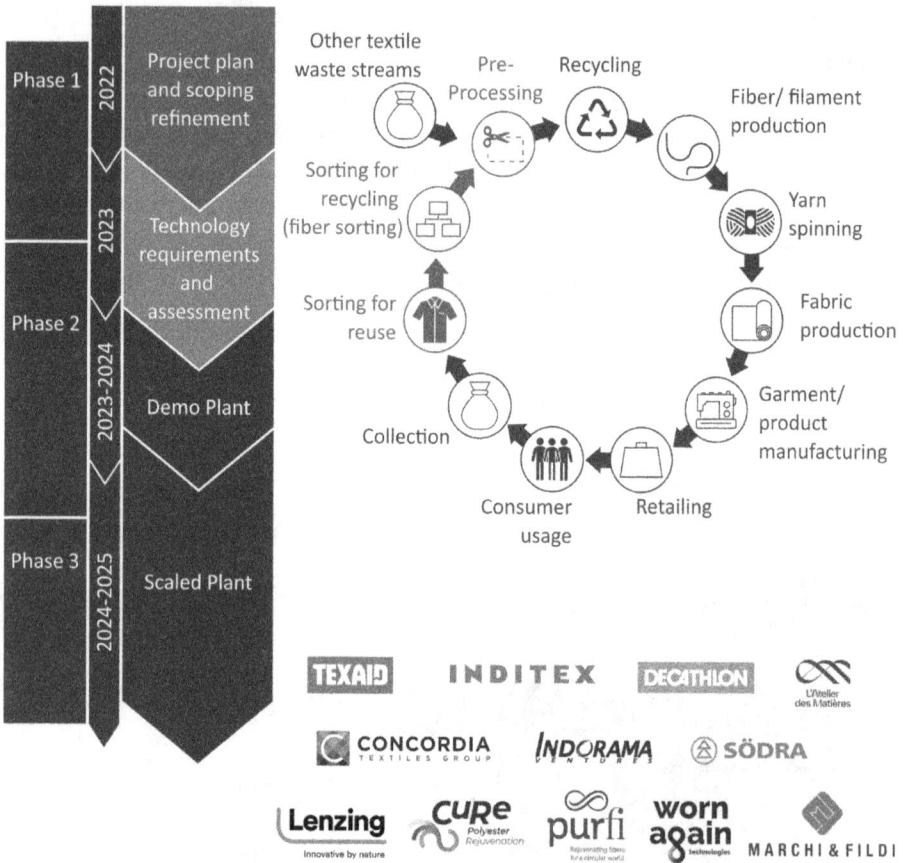

Figure 3.23 Project: "Turning Textile Waste into Feedstock"

An example from 2022–2023 is the cooperation of major stakeholders from various areas of the textile value chain on the project "Turning Textile Waste into Feedstock" (Figure 3.23). The initial aim of the project is, to conduct a technology screening to achieve higher automation in sorting for reuse, sorting for recycling, and preparation for recycling. The focus is on increasing and scaling sorting quantities through partial automation and improving the quality of the streams intended for recycling. The technology assessment also considers further research projects, such as digital product passport and recyclable designs.

Upon successful completion of the project, the newly created sorting plant is intended to serve as a blueprint for further intra-European sorting plants to meet the raw material requirements for downstream textile recycling. Only through such initiatives and projects will a circular economy of textile waste be possible.

3.2.3.4 Conclusion

The current linear textile value chain must become more circular and sustainable in the coming years. This necessary change will be driven and enabled by regulations regarding minimum collection quotas for textiles, restrictions on the textile waste exports, a growing second-hand market and advancements in sorting and textile-to-textile recycling technologies. Although collection and sorting of textile waste appears simple at first glance, it is highly complex and important, as it forms the backbone a circular textile value chain. Different methods of collecting textile waste yield varying degrees of textile quality. Not all collection methods are equal. Collection via bins might be the simplest and most common collection method, but often results in lower quality due to cross-contamination. Other methods, such as curbside collection, yield higher quality textiles, but incur high collection costs because of high personnel costs. The collection rate of post-consumer textile waste is estimated to increase from about 30% in 2022 to 50 to 80% by 2030. This will create additional sorting demand for 50 to 120 new sorting plants. Currently most sorting is done manually, requiring a large workforce to manage the complex sorting process. This consists of seven main steps: incoming goods acceptance; unloading; first sorting; second sorting; quality sorting; and factory-sorting; preparation for recycling, packaging and distribution. Automation and artificial intelligence technologies will be needed to support the sorting process in the future. Increasing automation in sorting for reuse, sorting for recycling, and preparation for recycling is crucial for increasing the quantity and quality of sorting stream, ultimately making the textile value chain more circular in the future.

References for Section 3.2

[1] European Commission Press release, *Circular economy for textiles: Taking responsibility to reduce, reuse and recycle textile waste and boosting markets for used textiles*; The EU imposes separate textiles waste collection by 2025, *https://ec.europa.eu/commission/presscorner/api/files/document/ print/en/ip_23_3635/IP_23_3635_EN.pdf*, *https://www.europarl.europa.eu/doceo/document/E-9-2020-004882-ASW_EN.html*, Brussels, July 2023

[2] European Commission, *Proposal of Waste Framework Directive Revision – Textile Waste*, *https://ec.europa. eu/newsroom/env/items/803765*, last accessed 23.04.2024

[3] European Commission, *Waste Framework Directive*, *https://environment.ec.europa.eu/topics/waste-and-recycling/waste-framework-directive_en*, last accessed 23.04.2024

[4]   European Commission, *Revision of the EU waste framework directive-textiles and food waste*, https://www.europarl.europa.eu/legislative-train/theme-a-european-green-deal/file-revision-of-the-eu-waste-framework, last accessed 23.04.2023

[5]   European Union, *Re4circular: a new AI-based solution to improve the sorting and cataloguing of textile waste*, https://www.circulareconomy.europa.eu/platform/en/good-practices/re4circular-new-ai-based-solution-improve-sorting-and-cataloguing-textile-waste, last accessed 23.04.2024

[6]   Pehrsson, A., Köppe, G., Geldhäuser, S., Krichel, A., Cetin, M., Pohlmeyer, F., Müller, K., Breton, J., *TECHNOLOGY ASSESSMENT – REDEFINING TEXTILE WASTE SORTING*: Impulses and findings for the future of next-gen sorting facilities.

[7]   European Union, *Siptex: A pioneering textile sorting technology for increased circularity*, https://circulareconomy.europa.eu/platform/en/good-practices/siptex-pioneering-textile-sorting-technology-increased-circularity, last accessed 23.04.2024

[8]   Statista, *https://www.statista.com/statistics/1008524/secondhand-apparel-market-value-by-segment-worldwide/*

[9]   Willersdorf, S., Krueger, F., Estripeau, R., Gasc, M., Mardon, C., *The Consumers Behind Fashion's Growing Secondhand Market*, Boston Consulting Group, October 2020, *https://www.bcg.com/publications/2020/consumer-segments-behind-growing-secondhand-fashion-market*, last accessed 23.04.2024

[10] De Klerk, A., *Secondhand clothing market set to be twice the size of fast fashion by 2030*, Bazaar, June 2021, *https://www.harpersbazaar.com/uk/fashion/fashion-news/a36810362/secondhand-clothing-boom/*, last accessed 13.04.2024

[11] Industry Insight BLOG Press Release, *The rise of the resale second-hand market in fashion*, Brussels, December 2021, *https://www.cbcommerce.eu/blog/2021/12/08/the-rise-of-the-resale-second-hand-market/*

[12] Study-Fashion 2030: *Trend guide for the future of the fashion industry in Germany*, *https://kpmg.com/de/en/home/insights/2021/01/studie-fashion-2030-trend-guide-fuer-die-zukunft-der-modebranche-in-deutschland.html*, last accessed 23.04.2024

[13] McKinsey & Company, Janmark, J., Magnus, K.-H., Strand, M., Langguth, N., Hedrich, S., *Scaling textile recycling in Europe – turning waste into value* (2022).

[14] BASF and Inditex make a breakthrough in textile-to-textile recycling with loopamid, the first circular nylon 6 entirely based on textile waste, joint news release January 2024, *https://www.basf.com/global/en/media/news-releases/2024/01/p-24-109.html*, last accessed 23.04.2024

[15] "TES of the ReHubs" *EURATEX report*; European Waste framework directive, European Waste Shipment Directive; Scaling textile Waste in Europe – turning waste into value

[16] Statista, *https://www.statista.com/statistics/263154/worldwide-production-volume-of-textile-fibers-since-1975/*, last accessed 23.04.2024

[17] Materials Market Report 2023, Textile Exchange 2023; last accessed 23.04.2024

3.3 Accelerating and Scaling Innovation – Recycling in India

Priyanka Khanna, Fashion for Good, Amsterdam, Netherlands,
Laura Barbet, Institut für Textiltechnik of RWTH Aachen University, Germany

3.3.1 Closing the Loop for Textile Waste in India

India, with its high volume of post and pre-consumer waste, presents a unique opportunity in the global push toward circular textile production. This waste, if mapped and sorted effectively, has the potential to serve as a quality feedstock for recycling technologies, thereby transforming what would be discarded into valuable resources. The ability to reuse these materials aligns with the principles of a circular economy, wherein waste materials are reintegrated into the production cycle, reducing the need for virgin resources and minimizing environmental impact. Utilizing waste as a feedstock for recycling not only diverts it from landfills but also reduces the industry's carbon footprint. Quality control in the sorting process ensures that the recycled materials meet the standards required for production, maintaining the integrity and value of the textiles. Moreover, this practice can significantly promote a circular textile value chain within India. By closing the loop, we see a direct connection between the end-of-life products and the beginning of a new production cycle. This is not just about waste management; it is about redefining the system to be designed to restore.

The strategic positioning of India in the textile industry, coupled with initiatives to close the loop for textile waste, can lead to a major change in how we view and manage textile production globally.

3.3.2 Objectives of the Sorting for Circularity India Project

The "Sorting for Circularity India Project" spearheaded by Fashion for Good is a strategic initiative with well-defined objectives to usher in a new era of sustainable textile management in India. This project aims to address the growing issue of textile waste through a series of planned actions.

The first goal is to develop a comprehensive understanding of the current textile waste material flow in India. This includes quantifying and categorizing the waste, distinguishing between pre-consumer, post-consumer domestic, and post-consumer imported waste streams. By mapping out these flows, the project aims to reveal the volume, categories, and values of the waste, which is crucial for devising effective management strategies.

Identifying recent technologies that can assist in the mapping and sorting of both pre- and post-consumer waste is the second goal. The focus here is on technological solutions that can address the complexities of waste management, ensuring an efficient sorting process that can handle a variety of textile waste.

Piloting the solutions is the third goal and involves testing the identified technologies to confirm their effectiveness. These pilot programs serve as proof of concept, validating the feasibility of the technologies for wider application in the textile waste management process.

Lastly, the project aims to build a roadmap for the implementation of these solutions. This involves collaboration with brands, manufacturers, recyclers, and other stakeholders to create access to high-quality waste feedstocks for recyclers. The roadmap will serve as a guideline for integrating these solutions into the existing textile network, ensuring that high-quality recycled materials are available and accessible.

This project is not just about technological innovation, it is about creating a sustainable infrastructure that supports circularity in the textile industry, aligning with global sustainability goals and the pursuit of a zero-waste future.

3.3.3 Collaborative Framework of the Sorting for Circularity India Project

The Sorting for Circularity India Project by Fashion for Good epitomizes a collaborative effort, bringing together key actors across the value chain to revolutionize textile waste management in India. This initiative is a concerted effort to close the loop on textile waste, led by a network of stakeholders, each playing a vital role in the supply chain.

Fashion for Good acts as the orchestrator, coordinating the symphony of participants ranging from brands such as adidas, Levi Strauss & Co., and H&M, to supply chain partners like Arvind and Welspun India. These brands and supply chain partners play a key role in the project, providing the necessary support to promote recycled materials and increase demand for sustainable textiles.

Key investors such as Laudes Foundation and IDH – The Sustainable Trade Initiative, are providing crucial capital and support, thereby driving the project forward. Their funding and resources are the bedrock upon which the project's ambitions are built, and enabling innovation and scaling of the circular economy model.

Innovators like Reverse Resources and Matoha contribute cutting-edge solutions and technologies that are vital to the sorting and recycling processes. Their innovations have the potential to transform waste into valuable resources, thereby reducing environmental impact and promoting sustainability.

The circle is completed by a myriad of other stakeholders, including manufacturers, recyclers, sorters, waste handlers, industry associations, ministries, and external consultants. Each entity brings expertise, resources, and commitment, driving the project toward its goal of establishing a closed-loop textile economy.

Together, these entities create a robust framework for change, demonstrating the power of partnership in addressing environmental challenges. This collaborative model is set to pave the way for sustainable practices, not just in India but as a blueprint for global adoption.

3.3.4 Three Project Deliverables

3.3.4.1 Textile-Waste Mapping Study

The "Wealth in Waste" study is the first of three crucial deliverables in the Sorting for Circularity India Project and it provides an in-depth look at India's potential to reintegrate textile waste back into the supply chain. Published in July 2022, the study is a cornerstone for understanding and addressing the textile waste challenge. The study encompasses several key areas of focus:

- Material Types: Cataloging the different types of textile materials present in the waste stream.

- Material Quantities: Measuring the volume of textile waste to assess the scale of the issue.

- Material Composition: Analyzing the makeup of the waste to determine the feasibility of recycling various materials.

- Existing Supply Chain: Evaluating the current supply chain to identify where improvements can be made.

- Waste Handlers: Understanding the roles and capacities of those currently managing textile waste.

- Sorters and Collectors: Looking at the existing infrastructure for sorting and collecting waste and how it can be optimized.

- Geographical Distribution: Mapping the regional spread of textile waste to inform targeted strategies.

- Challenges and Opportunities: Identifying the main challenges in managing textile waste and the opportunities for creating value from it.

By detailing these aspects, the study sets the stage for developing targeted interventions to improve waste management and recycling processes. It serves as a foundational resource for stakeholders across the industry to inform policy, investment, and operational decisions aimed at closing the loop in the textile life cycle.

The "Wealth in Waste" study conducted by Fashion for Good meticulously explores three primary waste streams, providing a holistic view of the textile waste landscape in India. These streams are categorized as:

- Pre-Consumer Waste: This stream encompasses waste generated during the manufacturing and production stages. It includes remnants from the cutting floor, defective products, and overstock – materials that have not yet reached the consumer.

- Domestic Post-Consumer Waste: This category includes waste accumulated after the consumer has used the product. It generally consists of discarded clothing and textiles from households within India, which presents opportunities for local recycling and reuse.

- Imported Waste: The third stream is the imported waste which India receives. These are post-consumer textiles that have been used in other countries and then imported into India for various purposes, including recycling.

By dissecting these streams, the study aims to assess the quantity and quality of waste available for recycling, understand the challenges of each category and identify potential interventions to optimize the recycling process. Recognizing the distinctions between these streams is essential for developing targeted strategies that can effectively channel waste back into the supply chain, thus supporting the circular economy model.

This study presents a quantitative analysis of textile waste streams in India and provides a clear overview of the scale and fate of textile waste. Approximately 7,793 kilotons (ktons) of textile waste are generated and imported into India each year. However, while a significant portion (59%) of this waste is reintegrated into the textile industry, only a small fraction re-enters the global supply chain, a fact which highlights a substantial opportunity for improvement in waste management and recycling practices. Of that 59%, the pre-consumer waste represents 3,265 ktons, which includes production leftovers and manufacturing waste. The post-consumer domestic waste consists of 3,944 ktons, which comes from within India itself. And finally, the imported waste amounts to 584 ktons, indicating global interdependencies in waste management, as illustrated in Figure 3.24.

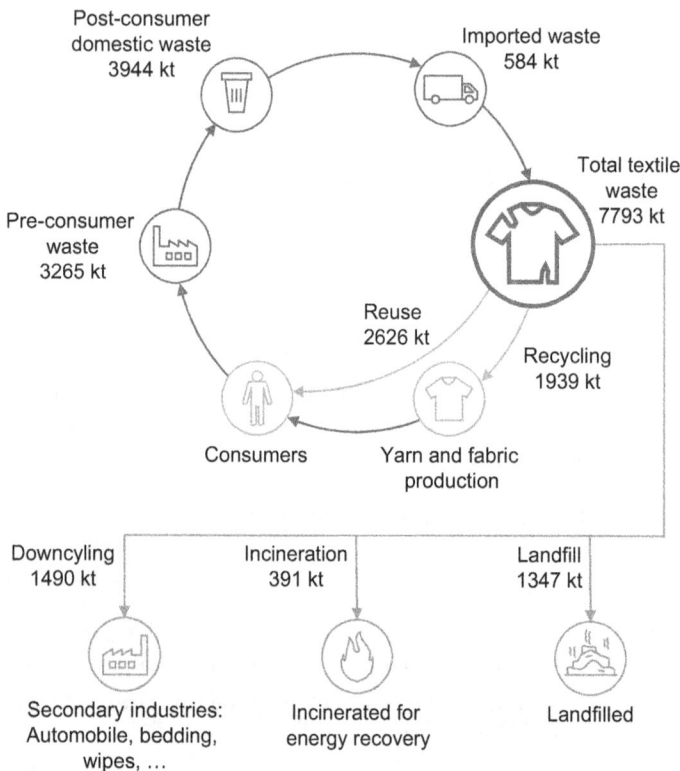

Figure 3.24 Quantitative analysis of textile waste streams in India

The remaining waste is either: downcycled, which represents 1,490 ktons (19%), often involving the conversion of textiles into products for secondary industries such as the automotive, bedding, and wipes industries. 5% is incinerated for energy recovery in plants such as brick kilns, small soap factories and boilers. And the final 17% ends up in landfill, a significant figure that represents the loss of potentially valuable materials from the supply chain.

This data underlines the critical need for enhancing the efficiency and scope of recycling and waste management processes in India. There is a clear indication that, while some systems are in place for managing textile waste, there is much room for innovation and development in circular practices to reduce landfill and incineration rates and increase the recycling and reuse of textiles.

The composition of textile waste is a crucial aspect in determining the recyclability and sustainability of textiles. The "Wealth in Waste" study reveals that approximately 61% of textile waste in India is composed of cotton and cotton blends, which are materials traditionally favored for their comfort and natural origins. However, the study also points out a significant and growing trend in the accumulation of polyester and other synthetic blends, which currently constitute 19% of the total waste, as illustrated in Figure 3.25.

This detailed analysis informs stakeholders in the recycling and waste management sectors, enabling them to tailor their approaches to the specific types of waste available. It also demonstrates the necessity for innovation in recycling technologies that can efficiently handle the diversity of materials present in textile waste.

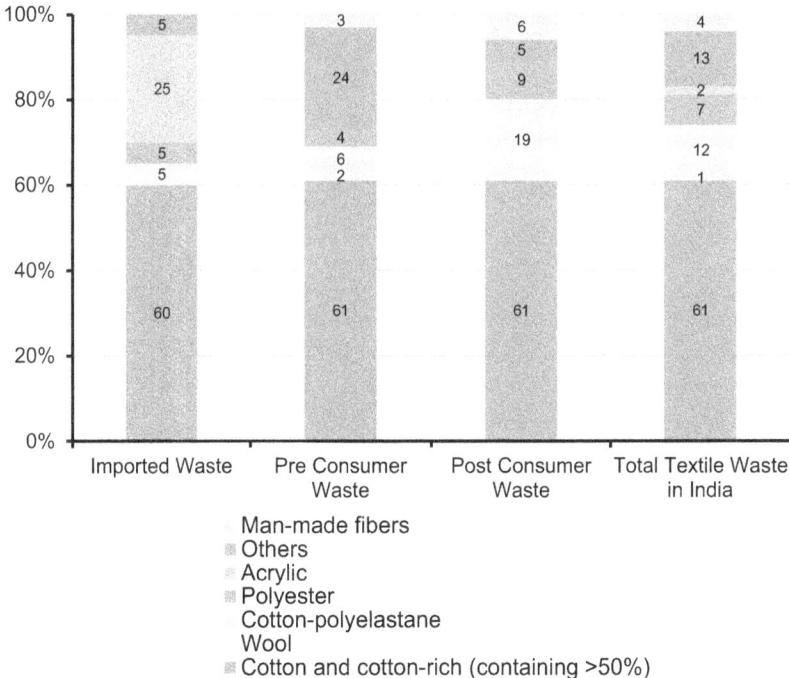

Figure 3.25 Composition analysis of textile waste in India

The network of textile waste flows within and into India, illustrated in Figure 3.26, converges on Panipat and Tirupur, cities recognized as major recycling hubs. The study identifies these hubs as the convergence points for the majority of textile waste, both domestically generated and imported. Panipat and Tirupur stand out as key players in the textile recycling industry, processing large quantities of waste and functioning as critical nodes in the global textile recycling supply chain.

The importation of second-hand clothing and mutilated rags, classified under HS Codes 6309 and 6310, is predominantly from six countries: the USA, Bangladesh, UAE, Canada, Japan, and South Korea. This importation demonstrates the global interconnectivity of the textile waste network and India's role within it.

Further illustrating the global textile cycle, the study notes that wearable second-hand clothing is often re-exported from India to various regions, including Africa, the European Union, and Southeast Asian countries. This re-exportation indicates not only the complexity of textile waste streams but also the potential for India to act as a significant redistributor of second-hand clothing.

This part of the study illustrates the importance of understanding the geographical dynamics of textile waste in the attempt to develop efficient recycling strategies and capitalize on the economic opportunities within these recycling hubs.

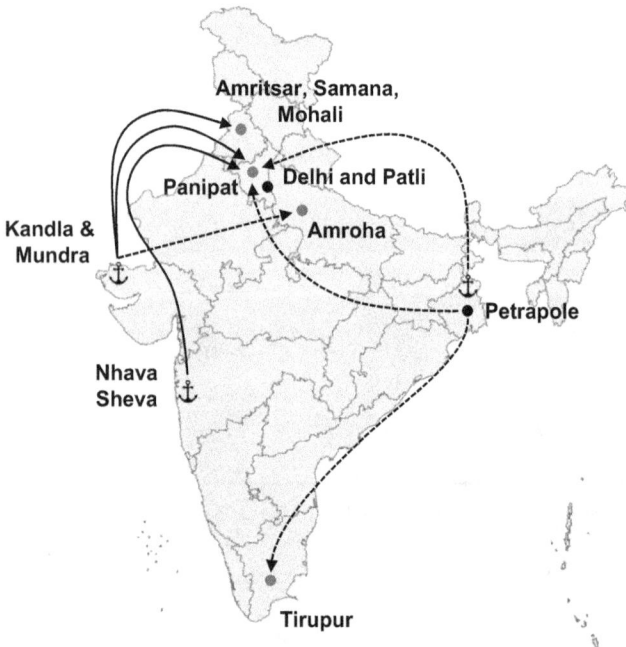

Figure 3.26 Textile waste imported into India

The "Wealth in Waste" study sheds light on India's textile waste value chain, acknowledging it as well-integrated but largely unorganized. The lack of standard practices across this value chain has been identified as a significant barrier to realizing the full potential of a circular economy in the textile sector.

The initial stage involves waste generation by manufacturers and consumers, which then enters the value chain. Waste is collected by various entities, including informal collectors and municipal organizations, highlighting the need for a more structured and efficient collection system. Aggregators play a crucial role in sorting and preparing the waste for further processing, but the unorganized nature of this stage can lead to inefficiencies. The waste is pre-processed, including by means of shredding, bleaching, and others that prepare the materials for recycling or reuse, and is then further processed to various outcomes, including reuse, recycling, downcycling, incineration for energy recovery, or landfilling.

The flow of textile waste through this value chain illustrates the complexity of managing such waste and the opportunities that exist for optimizing and standardizing practices. By doing so, India could significantly enhance the circularity of its textile industry, reducing environmental impact and creating economic value from waste.

The "Wealth in Waste" study incorporates a seven-level waste value hierarchy adapted from the EU Environmental Agency's framework under the headings Reuse, Repair/Reconditioning, High-Grade Recycling, Low-Grade Recycling, Downcycle, Incineration for Energy and Disposal. This hierarchy is employed to discern the perceived value of various waste types within the textile recycling industry, prioritizing waste management practices based on their environmental impact and resource efficiency.

The hierarchy acts as a guiding principle for sustainable waste management in the textile industry, encouraging the maximization of material utility and advocating for the reduction of textile waste's environmental footprint.

The study reveals that less than 50% of textile waste in India is currently realized for high value, indicating a substantial opportunity for improvement in the sector. High-value realization refers to the process of extracting the maximum possible value from waste materials, either through reuse, repair, or high-grade recycling, rather than downcycling, incineration, or landfilling.

Materials identified as having high-value potential include:

- Fabric Deadstock: Unused fabrics left over from manufacturing, which can be redirected to new production cycles.

- Re-wearable Clothing: Apparel that is still in good condition and can be worn again, either domestically or through export.

- Apparel Overproduction: Surplus clothing from overproduction, which can be sold in alternative markets or repurposed.

- White-Knitted 100% Cotton Waste: This pure cotton waste has a high recycling value, due to its single-material composition, and is therefore easier to recycle.

Solid colored cotton, blended materials with man-made cellulose fibers (MMCFs), and a wide range of printed textiles constitute a significant volume of the total waste generated in India's textile industry. These materials are prevalent because they are widely used in a wide range of clothing and textile products. While they are integral to the industry, their presence in the waste stream presents both a challenge and an opportunity for sustainable practices. The challenge lies in the complexity of recycling such diverse materials, which often requires different processes that will reclaim the fibers without compromising their quality. However, the opportunity is clear: if effectively recycled, these materials have the potential to significantly reduce the industry's environmental footprint by mitigating the need for virgin materials.

Currently, the industry is intensifying its on developing recycling technologies that can handle these materials more efficiently. The limitations of current recycling technologies often lead to downcycling, where materials are converted into products of lower quality and functionality. To prevent this value loss, there is a push toward creating advanced recycling methods, particularly for MMCF blends, which can be more challenging to recycle because of their mixed composition. The goal is to achieve high-grade recycling, where the quality of the recycled fiber is comparable to virgin materials, thereby enabling their re-entry into the high-end textile market. This approach not only preserves the intrinsic value of the materials but also aligns with the growing consumer demand for sustainable products.

The industry's shift toward this goal is driven by the realization that a sustainable approach to managing textile waste is not just environmentally prudent but also economically viable. By closing the loop on pre-consumer waste, the industry can reduce its reliance on raw materials, lower production costs, and cater to the eco-conscious consumer. This transition requires all stakeholders, including brands, manufacturers, and policymakers, to make a concerted effort to invest in new technologies, foster innovation, and create supportive regulations that can scale up these advanced recycling processes.

The study emphasizes the need for improved sorting and processing systems to enhance the rate of high-value realization. By focusing on these high-potential materials, the textile industry in India can increase its contribution to the circular economy, reduce environmental impact, and generate economic benefits.

- Sustainability depends in part on moving waste up the hierarchy which is driven by three key drivers: Industry Momentum and Building Pressure: There is growing momentum within the textile industry to adopt more sustainable practices. This momentum is driven by consumer awareness, environmental advocacy and regulatory pressures. The industry is recognizing the need to minimize waste and conserve resources, and this is leading to a reevaluation of production and waste management practices.

- Advanced Recycling Technologies: The development and impending implementation of advanced recycling technologies are setting the stage for higher quality feedstock requirements. These technologies promise to transform the way textile waste is processed by enabling the recovery of high-quality fibers, thereby closing the loop on textile materials.

▨ Demand for Traceability: There is increasing demand for traceability in the supply chain to ensure the sustainability and ethical sourcing of materials. Traceability serves as an enabler for scaling up recycling operations by providing the information needed for tracking the origin and journey undergone by textiles, which, in turn, supports the verification of sustainable practices.

These factors demonstrate the necessity to elevate waste management practices to higher levels of the hierarchy, where the emphasis is on reuse, repair, and high-grade recycling. By moving waste up the hierarchy, the industry can improve material utilization, reduce environmental impact, and foster a circular economy.

India stands on the brink of becoming a global leader in circular textiles, given its unique potential. However, the realization of this potential hinges on addressing several critical bottlenecks. Despite the fact that huge quantities of waste are generated and imported, a lack of traceability is a major barrier. Establishing systems to track the journey of textiles from production to disposal can enhance sustainability and compliance with environmental standards. India's waste value chain is well-established but predominantly informal and unorganized, which can lead to inefficiencies and lost opportunities for value recovery. While there is a use case for virtually all types of waste, the study highlights that the value potential is not fully realized. This suggests the need for better sorting, processing and end-use applications that can maximize value. The recycling industry in India is resilient but currently lacks access to advanced technologies that can improve the quality and efficiency of recycled textiles. The introduction of new technologies presents an opportunity to move on to more sophisticated recycling methods. However, the acceptability and success of these technologies in the Indian context have still to be fully ascertained.

By addressing these challenges, India could position itself at the forefront of the circular economy, maximizing resource utilization and setting an example for sustainable textile management worldwide.

3.3.4.2 Pre-Consumer Pilot Learnings and Findings

The "Sorting for Circularity" project's second deliverable is the Pre-Consumer Waste Pilot, a crucial initiative aimed at establishing a closed-loop system for managing waste generated during the production phase. This pilot, completed in November 2022, focused on the initial stages of the textile life cycle, aiming to optimize the utilization of resources and minimize waste before it reaches the consumer.

This pilot program focuses on several key processes illustrated in Figure 3.27.

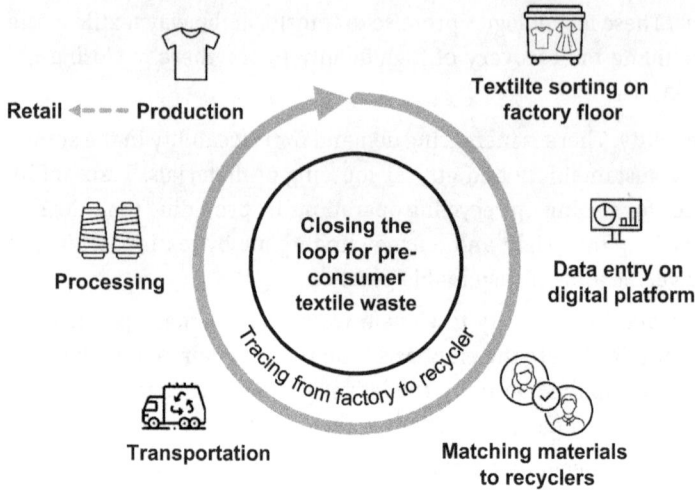

Figure 3.27 Closing the loop for pre-consumer textile waste

The initiative begins at the source, with the sorting of textile waste on the factory floor, which is an essential step in segregating materials for efficient recycling. Detailed record-keeping on digital platforms ensures the traceability of textile waste and allows for better tracking and management of waste streams. Efficient and responsible transportation systems are developed to move waste from production sites to recycling facilities, thereby minimizing the environmental impact of logistics. A critical component is the identification of recyclers who can process each type of material, as that ensures that waste is transformed back into valuable resources.

By closing the loop for pre-consumer textile waste, this pilot aims to demonstrate the feasibility of circular practices in the textile industry's value chain. It seeks to pave the way for systemic changes that reduce waste, preserve resources, and create a sustainable textile production model.

Figure 3.28 outlines a 360° approach to the managing of pre-consumer textile waste, illustrating the full cycle of waste, from its generation to its eventual recycling or other end-uses. This approach encompasses a collaborative effort between brands, manufacturers, and waste handlers, all coordinated through the Reverse Resources Platform.

Manufacturers generate waste during the production process that is then collected and categorized by the waste handlers. They mainly collect waste and play a crucial role in gathering this waste and directing it to appropriate recycling streams. The mechanically and chemically recycled material is then handled by the spinners, who produce coarse yarns or other textiles suitable for home textiles.

Other industrial uses for waste that cannot be reintroduced into the textile industry have been identified. This ensures that all waste generated is used to its fullest potential. The 360° approach underlines the importance of the Reverse Resources Platform as a coordinator for tracing and channeling waste efficiently. By connecting various stakeholders and standardizing the flow of materials, it enables the industry to maximize recycling, minimize waste, and promote a circular economy within the textile sector.

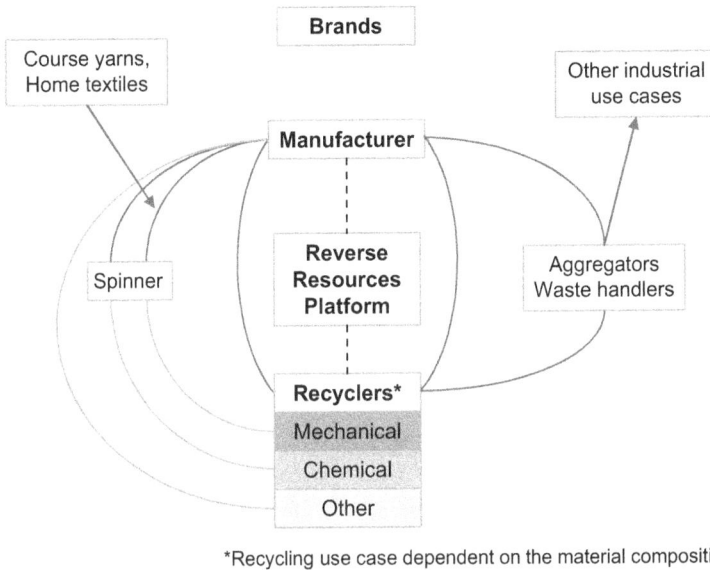

*Recycling use case dependent on the material composition

Figure 3.28 360° approach for pre-consumer textile waste

The "Sorting for Circularity" project recorded a significant milestone with the tracing of approximately 136 tons of pre-consumer textile waste during the project period, 84 tons of which was accounted for during the reporting period itself. This achievement marks a substantial step forward in the push toward a more sustainable and accountable textile industry. The waste, meticulously registered on a digital platform, provides a transparent overview of the waste volumes and their compositions – this demonstrates the effectiveness of the waste management strategies employed. The composition of the waste registered on the platform between June and November 2022 is summarized in Figure 3.29.

The detailed tracking of waste types – predominantly 100% cotton, polyester, cotton-poly blends – is illustrated in Figure 3.30 and reflects a concerted effort to understand and optimize the waste stream. With such data, the project has been able to ensure that a significant proportion of textile waste is directed toward appropriate recycling channels, rather than ending up in landfills or incinerators. The involvement of a diverse range of suppliers, as indicated by the data, showcases the collaborative nature of this initiative.

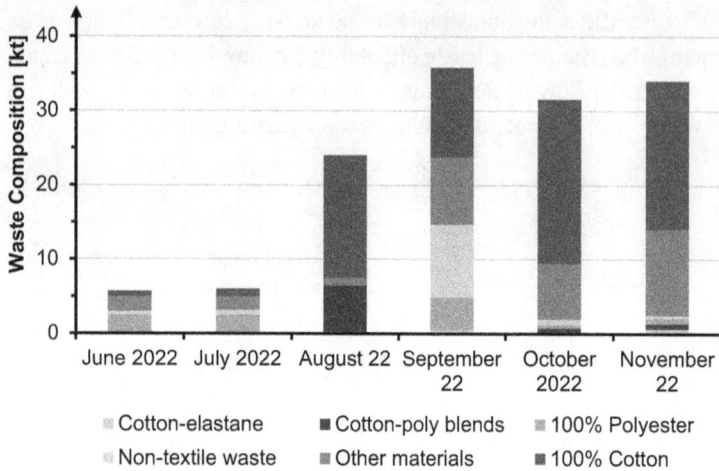

Figure 3.29 Composition of all waste registered on RR platform

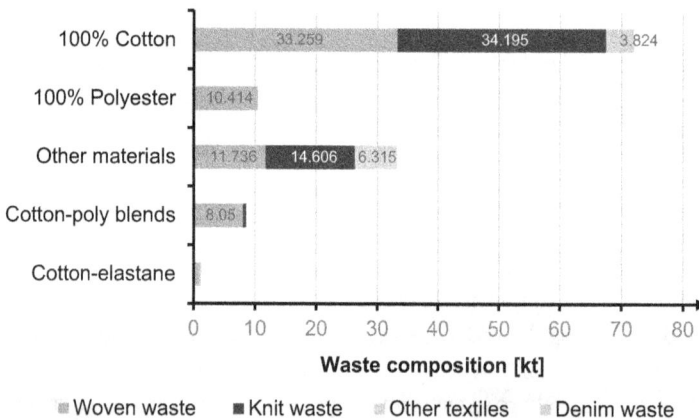

Figure 3.30 Key composition & materials types of textile waste

This project's success in tracing and documenting waste not only reinforces the value of transparency in waste management but also provides key insights into the types of waste that are most prevalent. These insights can inform future strategies for reducing waste at source and for developing new recycling technologies that can handle specific types of materials more efficiently. As the textile industry continues to grapple with sustainability challenges, initiatives like this offer a blueprint for harnessing technology and collaboration to create a more circular economy.

It achieved a significant victory by enhancing the visibility of sorted and traceable waste from its source to recyclers. This achievement demonstrates the effectiveness of the project's systems in mapping and managing the flow of pre-consumer waste. By tracking the journey of waste from manufacturers through waste handlers to recyclers, the project has established a transparent and efficient pathway for recycling.

Manufacturers such as Texport Industries, Aquarelle India, and MAF Clothing, have been instrumental in this process by not only segregating waste at the source but also participating actively in the waste tracing system. Waste handlers serve as the vital link, ensuring that the sorted waste reaches the appropriate recyclers. The recyclers, who range from those repurposing textiles into construction materials to those engaged in high-grade mechanical recycling, complete the cycle by transforming waste into new resources.

This streamlined process has not only led to the effective recycling of textile waste but also set a precedent for other industries to follow. It demonstrates the potential for creating a closed-loop system within the textile industry, where waste becomes a valuable input for new products, promoting sustainability and reducing environmental impact.

3.3.4.2.1 Challenges, Learnings and Long-Term Solutions

The "Sorting for Circularity" project, while pioneering in its approach, has encountered several challenges that have yielded valuable learnings and pointed to long-term solutions. The project's experience highlights four primary areas of focus. First, a fundamental challenge has been the lack of education and limited understanding of the market's shift toward sustainability. Addressing this requires comprehensive training and awareness programs that can enhance knowledge about the benefits of recycling and the importance of a circular economy. Next, the project has faced difficulties in forging agreements between manufacturers and recyclers. This operational challenge highlights the need for clear communication channels and mutually beneficial agreements that align the interests of all stakeholders involved in the recycling process. Furthermore, the economics of waste management have been a concern, with high supply-chain costs and dispersed volumes of waste. Streamlining operations and concentrating waste streams could present a more attractive business case as it would reduce costs and increase efficiency. Finally, the competition between early-stage high-value recycling technologies and established low-value technologies presents a technological challenge. Prioritizing investment in advanced technologies and supporting innovation can shift the industry toward more sustainable practices.

These insights pave the way for strategic initiatives that can overcome these challenges and foster an environment where sustainable textile recycling is not only possible but also profitable and widespread.

3.3.4.3 Post-Consumer Waste Pilot

The third crucial deliverable of the "Sorting for Circularity" project is the Post-Consumer Waste Pilot, which was scheduled from December 2022 to June 2023, but then took the baton and extended the initiative's reach into the post-consumer phase. This phase is critical as it deals with waste generated after the consumer has used the product, such waste often presenting more significant challenges in collection, sorting, and recycling.

This pilot project aims to create a sustainable network for dealing with consumer-level textile waste. It seeks to complete the circular journey of textiles by ensuring that garments and textiles discarded by consumers are collected, sorted, and prepared for a new life cycle, rather than ending up as landfill.

The pilot project hinges on several key actions, illustrated in Figure 3.31. First, post-consumer textiles are gathered from both domestic and imported sources, with the latter highlighting the global nature of textile consumption and waste. Then, a combination of manual and automated sorting methods is employed to efficiently categorize textiles on the basis of their material composition and recyclability. Afterward, the sorted textiles are directed to appropriate recycling channels, with some going to chemical recyclers, who can break down materials at the molecular level, and others to mechanical recyclers, who reprocess textiles into new fibers or products.

Secondary Use: Textiles that cannot be recycled are identified for secondary uses. This ensures that every piece of waste is utilized to its fullest potential.

Digital Integration: A digital marketplace plays a vital role in the pilot, allowing for the upload of data regarding waste streams and enabling the purchase and sale of materials. This adds transparency and efficiency to the process.

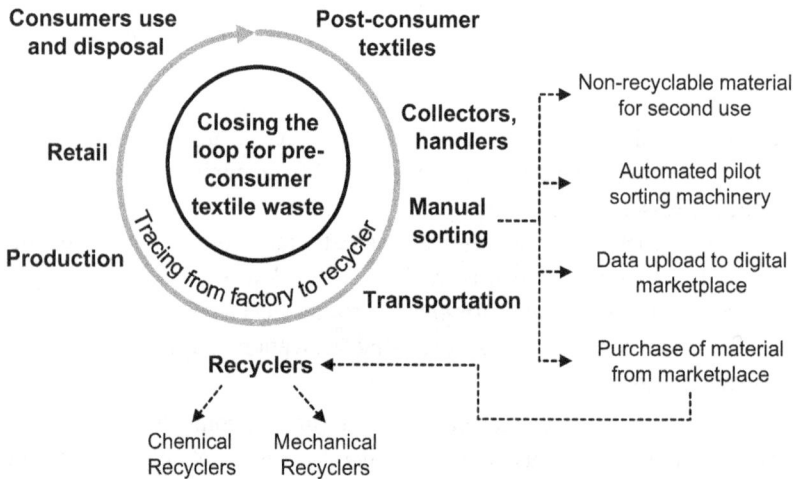

Figure 3.31 Post-consumer pilot

This comprehensive approach not only mitigates the environmental impact of post-consumer textile waste but also serves as a scalable model that can be replicated across different regions and industries.

3.3.4.4 Sealing the Cycle of Sustainability in Textiles in India

Following the successful achievement of the pilot phase, 2023 marked the beginning of the implementation stage, during which the validated technologies and insights were scaled and integrated into broader industry practices. This transition reflected the project's structured and methodical approach to closing the loop on textile waste in India.

Each phase of the project built upon the last, continually making progress toward a closed-loop system where textile waste is no longer treated as an end-of-life product, but as a valuable input for the creation of new materials. The project has finally demonstrated India's strong potential to collect, sort, and pre-process significant volumes of high-quality textile waste suitable for recycling. This is an essential step toward establishing a circular textile value chain.

Furthermore, the findings highlighted India's competitive position in Asia in terms of feedstock costs. However, the project also underscored the critical need for infrastructure development, particularly in collection and sorting systems. To unlock viable business models for sorting operations, it will be essential to establish robust collection, sorting, and resale channels. To achieve scalability, targeted funding and supportive policy frameworks will be crucial.

References for Section 3.3

Fashion for good. (2022). Retrieved from *https://fashionforgood.com/*

4
Recycling Strategies and Concepts

4.1 Mechanical Recycling–Background

Stefan Schlichter, Amon Krichel, Technische Hochschule Augsburg
(University of Applied Sciences), Germany

4.1.1 Principles

Mechanical recycling is one of the well-established methods for processing secondary raw materials and has been in use for many years. The basic principle is based on exposing and opening the textile, typically present as yarn composites or fabric, to gradually expose the fibers. These fibers can then be reused in the production of yarn or fabric as secondary fiber material.

In the waste hierarchy, which is considered binding for waste recovery by almost all governments, mechanical recycling falls into the category of material recycling. As illustrated in Figure 4.1, it is directly behind the reuse options in terms of preference. Mechanical recycling is characterized as an energy- and resource-efficient process. Like thermomechanical recycling, it is generally preferred to feedstock recycling (often referred to as chemical recycling), due to its better cost-effectiveness and energy efficiency [1]. Additionally, no additional substances need to be used for material separation and recycling by its nature, a fact which typically provides an ecological advantage as well.

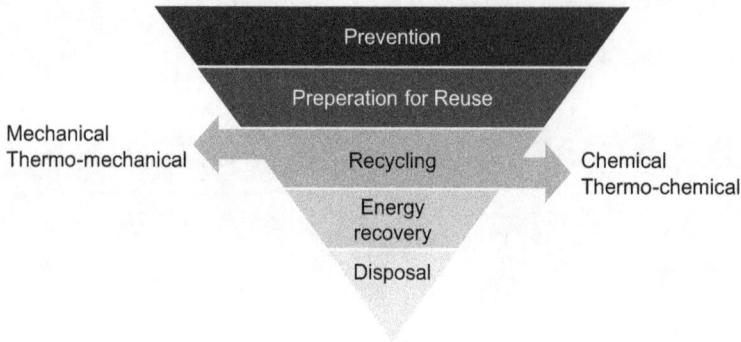

Figure 4.1 The top priority in the waste hierarchy is waste prevention

4.1.2 Workflow of Mechanical Recycling

Figure 4.2 illustrates the textile cycle as applied to mechanical recycling. After collection and sorting, the used textile is initially prepared for the actual fiber-opening process. In the subsequent processing stage, it undergoes tearing until it is opened to reveal the fiber.

The fibers processed by a tearing machine can typically be directly introduced into the traditional processes of spinning or nonwoven processing without further treatment. Processing into new textiles can occur as a pure recycled product or blended with other fibers. A significant advantage is that the integration of mechanical recycling into the existing process of yarn or nonwoven fabric processing is relatively straightforward from a plant layout perspective. However, adjustments regarding blend quality, color and achievable end product properties need to be made in order that the requirements of the final product may be met.

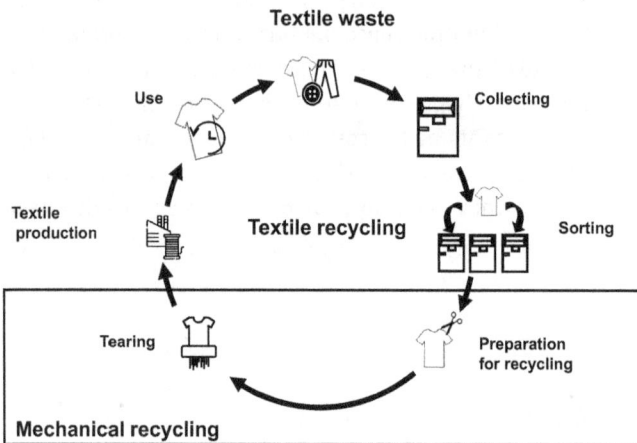

Figure 4.2 The mechanical textile recycling loop

4.1.3 The Process of Mechanical Recycling

As described above, mechanical recycling involves gradually reducing the size of, and opening, used textiles to expose the fibers for the subsequent processing stage. The process of mechanical recycling can be broadly divided into the preparation for recycling and the actual final opening process of tearing.

According to a study evaluating various recycling technologies [2], the status of the process can be assessed based on key criteria as shown in Figure 4.3:

	Mechanical Recycling
Output	fibers
Quality	-
Energy efficency	+
Cost advantage	+
Technology readiness level (TRL)	8
Requirements for sorting	~

*rating compared to other recycling technologies

Figure 4.3
Evaluation of mechanical recycling by key parameter [2]

Mechanical recycling offers clear advantages in almost all successful recycling categories. In particular, the efficient use of the energy needed for the process generates unmistakable economic benefits compared with other recycling technologies. In the current assessment, the already high level of technology readiness plays a significant role and is expected to provide a substantial advantage in the selection of potential recycling methods for the next few years. The main drawback compared with other recycling methods is that the fiber material undergoes significant stress through the tearing process, the outcome of which is lower quality of the torn material compared with the original. Current efforts are focused on optimizing this aspect to minimize the impact of this disadvantage.

4.1.3.1 Preparation for Recycling

Preparation for recycling entails ensuring that the textiles intended for the tearing machine are chosen on the basis of material type and transfer of the textiles in a form that lends itself to tearing. The preparation steps encompass the following sub-operations:

Material Selection

Given that the tearing process subjects the fibers to considerable stress, the achievable quality of the final product must be considered during pre-selection of the material.

Additionally, as the material mix generally cannot be influenced by the tearing process itself, the processing of mixed textiles is more challenging, even though the tearing process can, in principle, handle different materials. It is worth mentioning that mechanical recycling processes can, in general, be applied to all available types of fiber materials.

Sorting

In the sorting of used textiles for mechanical recycling, the pre-selection of fiber lengths present in the used textile is crucial, as it strongly influences the achievable quality level of the desired end product.

Reducing Size and Removal of Contaminants

The most important step in preparing for mechanical recycling is to shred the used textiles to a meaningful patch size, as only appropriately sized patches will lead to a continuous and consistent tearing quality. Various cutting devices are used for this purpose, with guillotine cutters, typically equipped with one or two consecutively arranged cutting blades, being the most common alongside circular cutters. Additionally, before the tearing process, larger contaminants such as zippers and buttons should be separated from the textile, as the removal of these contaminants is associated with high fiber stress and subjects the working parts of the tearing machine to significant strain and wear. Moreover, the separation of contaminants, double layers, and seams minimizes the stress on the material in the tearing machine, helping to maintain a high achievable fiber length. Special cutting devices (trimmers) are used for this purpose in the production line.

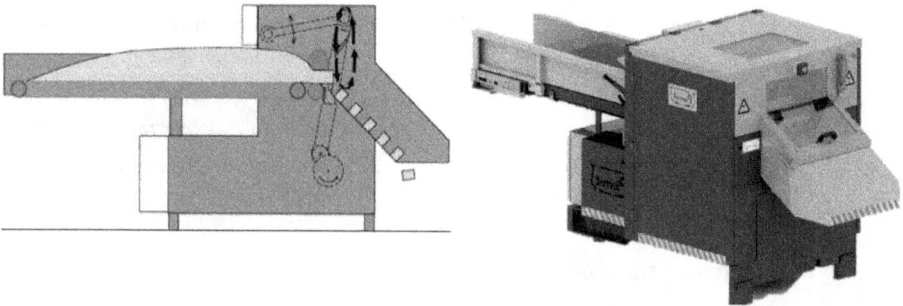

Figure 4.4 Preparation (guillotine cutting device) [4]

Creation of a Feeding Arrangement for the Tearing Machine

The textiles prepared as described above need to be processed into a continuous arrangement of patches in a suitable form to ensure uniform production at the tearing machine. A common approach is to compress bales after preparation, which then, at the tearing machine and upstream mixing devices, ensures a homogeneous feed for subsequent processing.

4.1.3.2 Tearing Machine

Figure 4.5 shows a typical tearing machine. The essential structure of the machine is already evident in the illustration, where the input material is supplied to the machine via conveyor belts. Subsequently, the material passes through a sequence of feed devices, tearing drums equipped with opening units, associated working elements, and material discharge through material-air separation (such as a condenser).

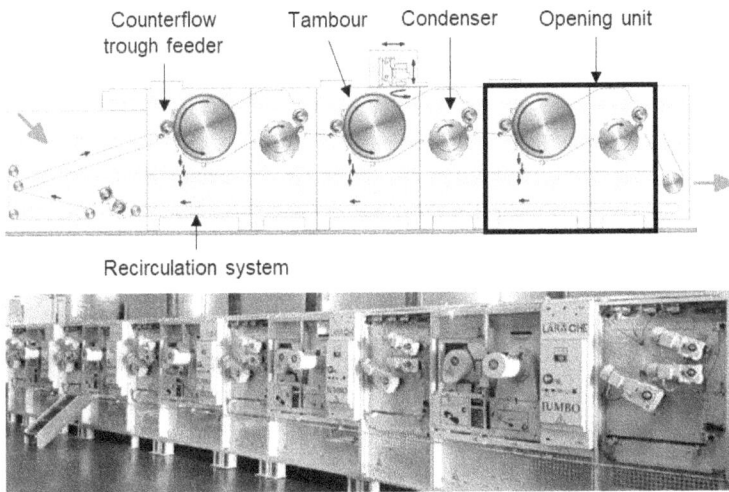

Figure 4.5 Cross-section of Andritz-La Roche-Jumbo machine [5]

Figure 4.6 shows the working elements of a tearing machine and lists various designs that play a role in the machine's performance and quality. The following overview reveals that tearing machines generally employ principles similar to those used in fiber opening or cleaning in the context of cotton processing [3].

Feeding

The material to be treated is fed to the machine through the feeding device, where, analogous to spinning preparation, either pure roller feeds or rollers with feed trays are used. To achieve the opening effect, forces need to be realized, and so preference is given to roller feeds or configurations with rollers/feed trays operating on the principle of counterflow feeding. To limit the risk of fiber damage associated with intensive opening, in addition to the mentioned traditional feeding configurations, arrangements with rollers/feed trays operating on the principle of parallel feeding are now also used. In these configurations, more intensive opening steps on the main drum must follow.

Main Cylinder

The main working element of a tearing machine is represented by the tambour or main cylinder, as the essential working elements that perform the opening work are arranged on the circumference of the main cylinder. The cylinder can be equipped with coarse porcupine beaters, pins, or sawtooth wire. Since multiple working stages are typically implemented in a tearing machine, the fittings are gradually made finer in the material flow.

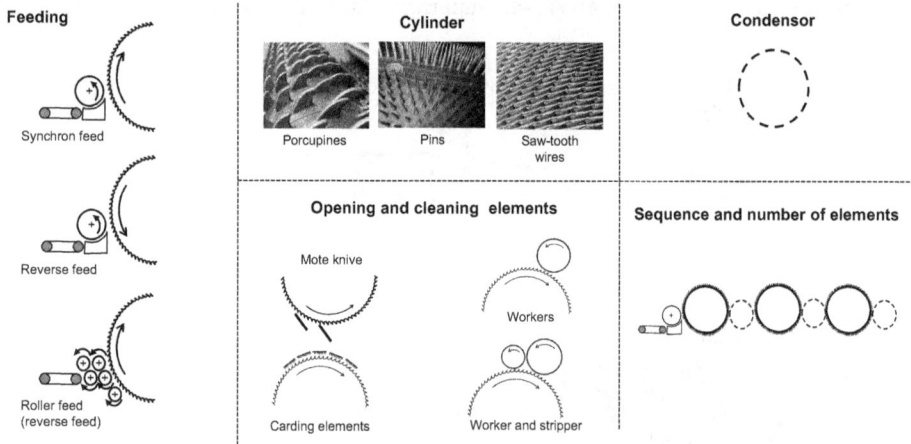

Figure 4.6 Main working elements of a tearing machine

Opening and Cleaning Elements

Arranged around the main cylinder in the material flow are additional working elements to ensure further opening of the fiber material and the removal of undesirable foreign particles after opening by the feeding zone. Traditionally, mote knives are arranged around the drum to ensure the ejection of non-dissolvable material components and foreign objects from the process. Recently, worker or worker/stripper arrangements have also been employed in tearing machines, similar to those known in long-staple processing on carding machines, such as carding willow or roller cards. In principle, the use of stationary carding elements is possible, but this application is limited due to the generated heat and the high stresses, requiring combinations only with other working elements such as mote knives and workers.

Sequence and Number of Elements

As the desired tearing result is often not achieved in just one processing stage, it is typical to have multiple tearing stages, each consisting of a feeding unit, a drum and corresponding opening and cleaning elements, arranged consecutively. The number and type of working elements used then determine the achievable tearing quality in terms of purity and acceptable fiber damage.

Condenser

After each drum, a device called a condenser is used to transport the material or to transfer it to the next tearing stage. This unit ensures the separation of the material from the accompanying air volume and simultaneously provides effective dust removal, which is necessary for eliminating high quantities of short fibers and dust. Typical embodiments of current tearing machines, illustrating some of the described combinations as per Figure 4.6, are shown in Figure 4.5 and Figure 4.7.

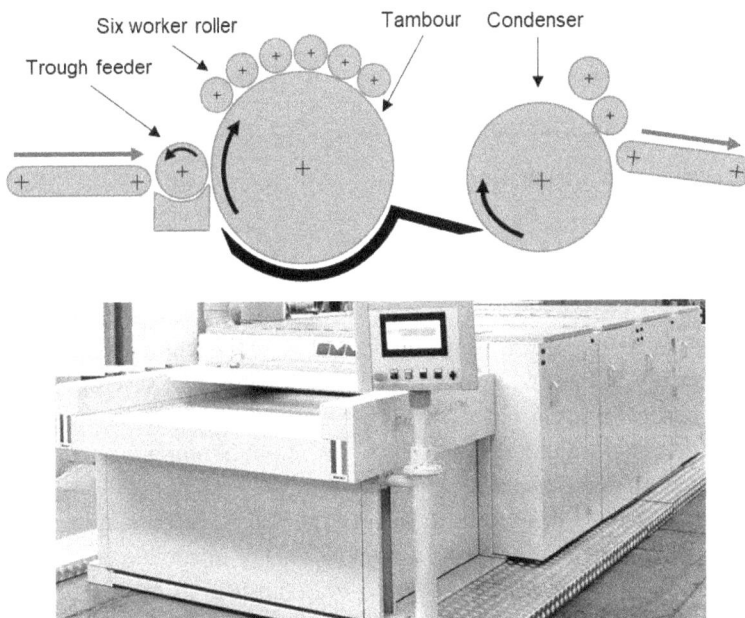

Figure 4.7 Cross section of Ommi Srl – Recoline machine [6]

4.1.3.3 Plant Configuration for Mechanical Recycling

Depending on the processed material, the desired product range, and the specified quality, tearing systems differ according to the specifications described above. Figure 4.8 illustrates an overall plant layout where all necessary units for preparation and the actual tearing process are presented in a typical arrangement. Typically incorporated between the actual preparation and tearing processes are a mixing system and/or baling for pressing to ensure the quality and continuity of material feed. The depicted arrangement shows a tearing machine configuration with 6 drums.

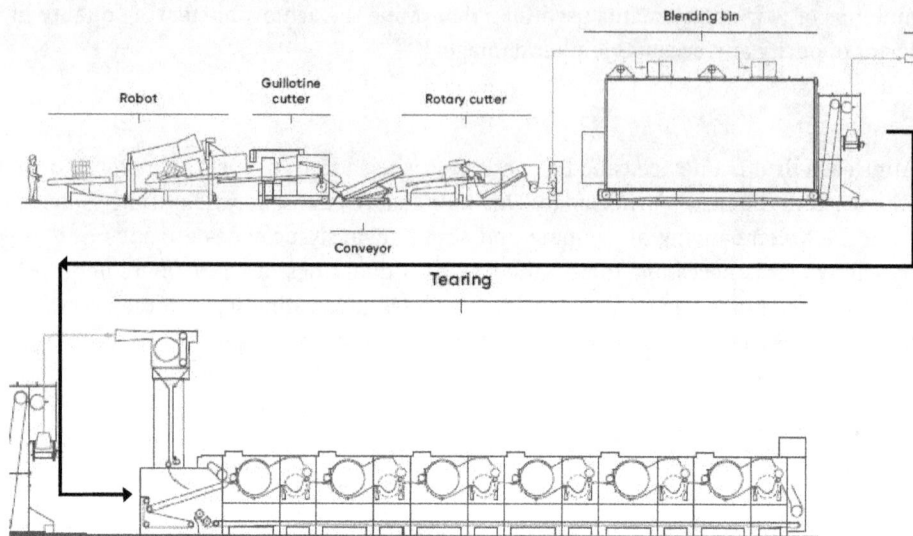

Figure 4.8 Plant configuration of a tearing line [7]

4.1.4 Settings for Mechanical Recycling

In addition to the described arrangements of process stages and machines, the attainable result is significantly influenced by the input material and the chosen process settings. The central task of an optimized tearing process is to adjust the fiber length or fiber length distribution, the content of short fibers, the output of non-processable fibers, and the degree of opening, especially in relation to yarn remnants that so far could not be opened. This adjustment aims to achieve the highest possible quality of the processed material concerning the desired product quality in combination with the subsequent processing stage in spinning or nonwoven fabric processing. While in the past, torn fiber materials from mechanical recycling were primarily used for lower-quality products, the current goal of research is to achieve higher end product qualities. Understandably, this is only possible when the chosen settings in mechanical recycling are ideally tailored between upstream sorting and subsequent processing in spinning or nonwoven fabric processing. Monitoring the development of fiber lengths across various process stages will be crucial to producing high-quality end products from used textiles.

Another challenge for mechanical recycling is the enabling of new processing strategies. Cascade recycling is particularly noteworthy. It aims to mitigate the unavoidable loss in quality, in the form of reduced fiber length, by applying the product in a slightly lower-quality product category. Furthermore, due to the need to optimize fiber length, the addition of virgin fibers is employed as an effective measure for achieving high qualities, especially in the field of mechanical recycling.

References for Section 4.1

[1] McKinsey & Company, Scaling textile recycling in Europe–turning waste into value: Textile recycling can turn Europe's textile waste into value and build a sustainable and profitable new industry. (2022).

[2] Cherdron, B., Krichel, A., Schlichter, S., Naji, F., Reinelt, B., Schönemann, M., Nordsieck, H., Kroban, M., Tronecker, D., Dietz, W., Albert, A., Rommel, W., Ökonomische Potenziale des Textilrecyclings und der Wasserstofferzeugung aus Textilabfällen in Bayern, Studie des ITA Augsburg gGmbH und bifa Umweltinstitut, 2023

[3] Gries, T., Veit, D., Wulfhorst, B., Textile Fertigungsverfahren, 3. Auflage, Hanser Fachbuchverlag 2018

[4] Pierret s.r.l. Corbion, Guillotine Schneidmaschine für Weiche Materialien – Datenblatt Pierret S45 Corbion Belgium

[5] Andritz AG, Graz, Andritz&Laroche recycling brochure, *https://www.andritz.com/products-en/nonwoven-textile/textile-recycling/jumbo-and-jumbo-exe*, Accessed on 21.12.2023

[6] Ommi S.r.l Prato, RecoLine, *https://www.ommi.it/2021/06/30/recycling/* Accessed on 21.12.2023

[7] Andritz AG, Graz, Andritz&Laroche recycling brochure, *https://www.andritz.com/products-en/nonwoven-textile/textile-recycling/jumbo-line*, Accessed on 21.12.2023

4.2 Mechanical Recycling: The Säntis Textiles RCO100 System

Annabelle Hutter, Säntis Textiles AG, Bühler, Switzerland

4.2.1 Introduction

In a world increasingly concerned with environmental sustainability, and facing ever-increasing EU regulations, the textile industry stands at the forefront of the movement toward more circular practices. Säntis Textiles, a Swiss multi-generational textile company with a rich history in textile engineering and machinery, has emerged as a pioneer in the field of textile recycling. Leveraging decades of knowledge and expertise, Säntis Textiles has developed the RCO100 system, a groundbreaking innovation in textile recycling technology.

4.2.2 Origins and Development

The journey toward the creation of the RCO100 system began in 2016 when Säntis Textiles embarked on a mission to build a mechanical recycling system for 100% recycled cotton yarns and fabrics. The driving force behind this endeavor was the company's commitment to sustainability and its recognition of the need for more environmentally friendly solutions within the textile industry. Partnering with PVH Corp, a global leader in the fashion industry, Säntis Textiles set out to achieve a lofty goal: to produce high-quality recycled cotton yarns and fabrics without compromising on fiber length and strength.

By the end of 2018, Säntis Textiles had reached a significant milestone, successfully spinning open-end yarns up to Ne20 using 100% pre-consumer industrial waste. These yarns served as the foundation for a range of 100% recycled textile products, the first in the world, including denim, chino, canvas, and jersey fabrics. Collaboration with PVH Corp. culminated in the introduction of 100% recycled denim at a Paris fashion show in early 2019, showcasing the potential of the RCO100 system to revolutionize the fashion industry. In 2022, Säntis reached a new milestone in creating the world's first 100% recycled jeans, made from 50% post-consumer and 50% pre-consumer waste, with the post-consumer waste collected from in-store take-back programs in PVH retail points across the Netherlands.

4.2.3 Partnerships and Collaborations

Central to the success of the RCO100 system has been Säntis Textiles' strategic partnerships with industry leaders such as Kipas Textiles and Temsan in Turkey, the former a vertical manufacturer and the latter a global supplier of textile air conditioning equipment. The joint venture with Kipas led to the installation of a custom-built recycling line at their facility in Kahramanmaras, Turkey, where the RCO100 system has been producing approximately 300–350 tons of recycled cotton bales per month since 2021.

Temsan, renowned for its expertise in textile machinery, played a crucial role in the development and production of the RCO100 system. With components engineered in Switzerland, the RCO100 system represents a fusion of cutting-edge technology and meticulous craftsmanship. Säntis Textiles' decision to license the production, sales, and service of the RCO100 system to Temsan underscores their commitment to collaboration and innovation in the pursuit of sustainable textile production.

4.2.4 Key Features and Innovation

At the heart of the RCO100 system lies its revolutionary mechanical fiber-opening technology, designed to treat fibers gently and preserve their length and strength. Unlike traditional recycling systems, which often result in fiber degradation, due to excessive stress, the RCO100 system prioritizes fiber quality, ensuring that recycled cotton maintains its integrity throughout the recycling process.

The system's unique features include a multi-stage processing mechanism, modular design for flexibility in handling various cotton inputs, and advanced dust collection and filtration systems. Additionally, ongoing research and development efforts are focused on incorporating patented technologies for elastane extraction, direct can coiling, and traceability, to further enhance the system's capabilities and sustainability credentials.

Säntis Textiles has amassed unparalleled expertise in textile manufacturing. Drawing upon this wealth of knowledge, the company embarked on a mission to develop a mechanical recycling system for 100% recycled cotton yarns and fabrics. The goal was clear: to maintain the best fiber length and strength possible while minimizing environmental impact. Since 2016, Säntis Textiles has been at the forefront of research and development in textile recycling through its focus on cotton – a staple material in the textile industry.

4.2.5 The RCO100 Fiber

Fiber length plays a crucial role in determining the quality and versatility of yarns produced in the textile industry. An understanding of the output fiber length obtained through recycling processes is essential for optimizing yarn production. The implications of output fiber length on yarn manufacturing and the types of yarns that can be produced with varying fiber lengths are crucial.

In textile recycling, the output fiber length is a key parameter that directly influences the quality and characteristics of the final yarn. When recycling Turkish cotton, which typically has an 80% staple length of approximately 31–32 mm, the resulting output fiber length after recycling with the RCO100 system is approximately 25–26 mm. This reduction in staple length is attributed to the mechanical processing involved in recycling, which may result in some fiber breakage and shortening. Factors such as the input fiber length and the recycling process itself significantly impact the output fiber length. While the initial fiber length of the input material sets the baseline, the RCO100 recycling process may introduce some loss, typically ranging from 15% to 20%. However, adjustments in production rate can help mitigate this loss, allowing for the retention of longer fibers in the final output.

4.2.6 Yarn Production

With the RCO100 machine, the output fiber length obtained through recycling has profound implications for yarn production. Different types of yarn require specific fiber lengths in order that the desired characteristics such as strength, softness, and texture may be achieved. An understanding of the output fiber length enables textile manufacturers to tailor their yarn production processes accordingly. Open-end yarns, characterized by their cost-effectiveness and suitability for bulk production, can be spun from RCO100 fibers with staple lengths ranging from Ne5 to Ne20. These yarns are commonly used in various applications, including apparel, home textiles, and industrial fabrics. As the output fiber length obtained from recycled Turkish cotton falls within this range, it is suitable for producing open-end yarns.

In contrast, carded ring compact yarns, known for their superior strength, smoothness, and uniformity, require longer fibers for optimal results. RCO100 fibers allow for staple lengths ranging from Ne8 to Ne30, and these yarns are favored for high-quality textiles and premium apparel. While the output fiber length from recycled Turkish cotton may fall within this range, manufacturers may need to carefully control the production rate so as to retain longer fibers and ensure yarn quality. Furthermore, advancements in yarn technology have led to the development of carded ring compact core yarns, which incorporate recycled elastane for added elasticity and performance. These yarns, with staple lengths up to Ne20 and deniers of 40 and 70, offer enhanced comfort and flexibility, and are ideal for activewear, sportswear and other stretch fabrics.

Output fiber length plays a critical role in determining the types of yarn that can be produced through textile recycling processes. An understanding of the relationship between input fiber length, recycling methods and output fiber length is essential for optimizing yarn production and achieving desired product characteristics. By leveraging this knowledge, textile manufacturers can effectively utilize recycled materials and contribute to the sustainability of the industry while meeting consumer demands for high-quality, recycled textiles.

4.2.7 The RCO100 System

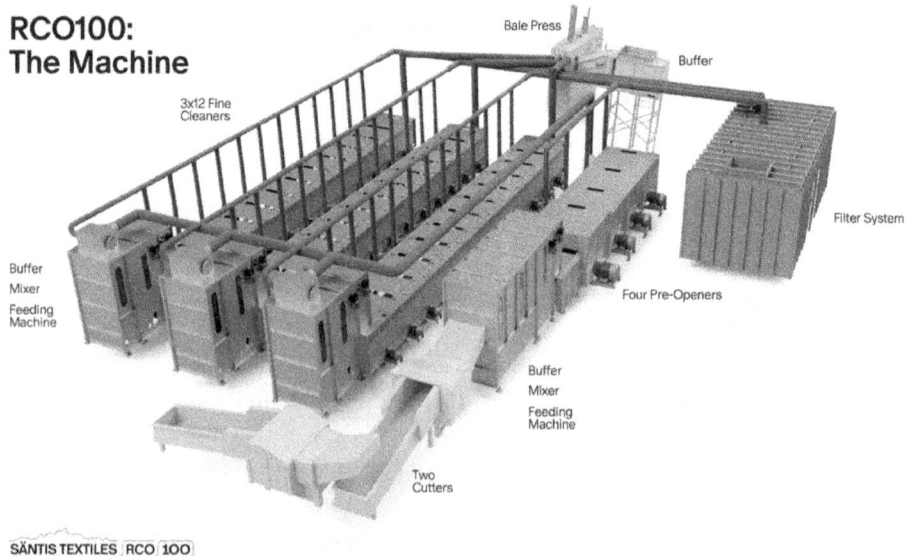

Figure 4.9 RCO100: The Machine

The RCO100 system boasts a myriad of unique features that distinguish it from conventional recycling methods (Figure 4.9). At its core lies a commitment to gentle fiber treatment to ensure maximum fiber length and strength. The system includes two conveyors and guillotine knives positioned at a 90° angle for optimal cutting results.

A sophisticated buffer unit controls the feeding to pre-opener cylinders, while spark and metal detection ensure operational safety. The pre-opener and fine-opener lines are equipped with advanced modules and cylinders, providing unmatched efficiency and performance.

By enabling the recycling of a wide range of staple fiber fabrics, including cotton, polyester, viscose, and more, the system reduces reliance on virgin materials and minimizes waste. The output fibers, with staple lengths of approximately 25–26mm, are suitable for spinning open-end yarns and carded ring compact yarns across various counts. Additionally, the system's ability to accommodate recycling elastane up to 4% further enhances its credentials. The system's output fibers can be spun into a wide range of yarns, including open-end yarns, carded ring compact yarns, and core yarns with recycled elastane. Looking ahead, Säntis Textiles envisions incorporating patented technologies for elastane extraction, direct can coiling, and traceability. This will further enhance the system's capabilities and sustainability credentials.

The RCO100 recycling system comprises several key components that have been meticulously designed to facilitate the recycling process:

- Cutters: Equipped with two guillotine cutters arranged at a 90° angle, the system ensures precise cutting of materials, enhancing efficiency and safety.

- Pre-Opener Section: This section consists of multiple pre-opener modules with various cylinder diameters and working widths, allowing for efficient initial processing of materials.

- Fine Opener Section: Comprising fine opener modules, this section further refines the materials, preparing them for subsequent stages of recycling.

- Buffer/Mixing Unit: Responsible for feeding materials to the fine opener section, the buffer/mixing unit ensures smooth material flow and optimal processing.

- Bale Press: The system features a double head bale press capable of producing multiple bales per hour, contributing to efficient material handling and storage.

- Filter Package: Incorporating pre-filter and rotary filter components, the system's filtration system effectively removes dust and debris, maintaining operational efficiency and air quality.

The efficiency and capabilities of the RCO100 recycling system are underscored by several factors:

- Production Capacities: With a total system capacity of 900 kg/h, the RCO100 system demonstrates high throughput, enabling the processing of large volumes of materials within a relatively short timeframe.

- Energy Efficiency: The system is designed for optimal energy utilization, with a total installed power of 405 kW. Inverter control for reducers and motors further enhances energy efficiency, minimizing operational costs and environmental impact.

■ Material Handling: The RCO100 system ensures gentle treatment of fibers, preserving their length and strength throughout the recycling process. This results in high-quality recycled materials that are suitable for various textile applications.

■ Safety Features: Incorporating metal detection and spark/fire protection systems for each module, the RCO100 system prioritizes safety in operation, mitigating potential hazards and ensuring a secure working environment for operators.

Figure 4.10 lists some additional technical specifications of the RCO100 system.

Cutters	Pre-Opener Section	Fine Opener Section	Buffer/Mixing Unit	Production Capacities	Equipment Specifications	Bale Press	Filter Package
Two guillotine cutters arranged at a 90° angle	Stage 1: 1. 1 x Pre-opener module 2. Cylinder diameter: 1000 mm 3. Working width: 1500 mm	Stage 1: 1 x Fine opener module 1. Cylinder diameter: 400 mm 2. Working width: 1500 mm	Feeds material to fine openers	Pre-opener line capacity: 900 kg/h	Total installed power: 405 kW	Double head bale press capable of producing 6 bales per hour	Consists of cabinet-type pre-filter and rotary filter
Equipped with conveyor belts, feeding funnels, and buffer/mixing unit	Stage 2: 3 x Pre-opener modules 1. Cylinder diameter: 700 mm 2. Working width: 1500 mm	Stage 2: 11 x Fine opener modules 1. Cylinder diameter: 250 mm 2. Working width: 1500 mm	Features chute system for smooth material flow	Fine opener line capacity: 300 kg/h (3 x 300 kg/h = 900 kg/h total fine opener capacity)	System control: Lenze	Each bale weighs between 190-200 kg	Capacity: Up to 70,000 m3/h
Spark and metal detection features for enhanced safety		Standard system comprises 3 fine opener lines with a total of 36 opening points		Total system capacity: 900 kg/h (based on input material, figures based on yarn waste)	Inverter control for all reducers and motors	Total capacity: 1200 kg/h	
					Reducers: Lenze		
					Electric Motors: Siemens IE3		
					Inverters: Lenze		
					Metal detection and spark/fire protection system for each module		

Figure 4.10   Additional technical data of the RCO100 system

4.2.8 Summary

Säntis Textiles, a Swiss-based textile company, developed the ROC100 System, an innovative mechanical recycling system for pre- and post-consumer waste. At the beginning stood a partnership with PVH, a global leader in the fashion industry, who was looking for more sustainable solutions. Crucial for the successful development of the system was the collaboration with Kipas Textiles, an integrated textile company in Türkiye and Temsan, a Turkish supplier of textile air conditioning equipment. The main focus of the ROC100 System is the gentle recycling of waste with the objective of preserving as much length and strength as possible of the original fibers. The system makes it possible to produce recycled cotton fibers of 25–26 mm in length if the virgin fibers had originally a length of 31–32 mm. It can recycle not only cotton but also other staple fibers like polyester, viscose, and others. Furthermore, it can also recycle fabric waste with an elastane content of up to 4%.

4.3 Textile Circular Solutions: Thermomechanical Recycling

Hafiz Kaleem, Javier Vera Sorroche, Centre Européen des Textiles Innovants (CETI), France

4.3.1 Introduction

Textile waste is found at various stages of the textile value chain, and recycling this waste can save a significant amount of material resources. Currently, post-industrial waste is predominantly recycled, whereas distribution and post-consumer waste are recycled in limited quantities, a fact which poses a substantial challenge to sustainable textile practices.

Textile recycling presents a complex challenge, with varying degrees of ease and difficulty that depend on the type of textiles involved. Distinct textile waste streams exist and are categorized by the type of textiles, such as apparel, home textiles, and technical textiles. Among these, apparel and home textiles are relatively easy to recycle, whereas technical or industrial textiles pose significant recycling challenges, due to their intricate construction.

Furthermore, the duration of textile use varies, with some textiles being used briefly, such as face masks and non-woven gowns, and others serving longer-term purposes, like curtains and firefighter uniforms. The recycling standards and methods for different types of textiles are likely to differ accordingly, reflecting these variations in use and complexity. It is worth noting that industrial or technical textiles present the greatest recycling challenges and are currently among the least recycled textiles.

Closed-loop textile recycling, often referred to as the textile circular economy, is a sustainable and environmentally aware approach to textile production and consumption. In a closed-loop system, textiles are designed, produced, used, and recycled with the goal of minimizing waste, conserving resources and reducing the environmental impact of the textile industry. This approach is part of a broader movement toward sustainable and eco-friendly practices in the fashion and textile industries [1].

Figure 4.11 presents the key steps included in closed-loop textile recycling.

Textile waste collection, which accounts for around 30% of textiles placed on the market annually, is primarily conducted through separate collection systems, often involving designated collection points such as retail stores. The textiles collected separately are typically directed to global second-hand markets for reuse. However, among the textiles collected, a portion cannot be reutilized, due to factors such as heavy use, cleanliness issues, damage, or low quality, these factors classifying them as post-consumer textile waste [2].

Figure 4.11 Closed-loop textile recycling

The volume of collected textiles is on the rise, largely driven by the expanding fashion industry. Proper sorting of these textiles is essential for reuse or recycling. Currently, manual sorting predominates and it significantly influences the financial aspects of textile sorting activities. However, with the increasing volume of post-consumer textile waste, manual sorting is becoming less efficient for textile recycling. Moreover, efficient textile recycling requires the precise identification of feedstock and a pollution-free environment in order that high-quality recycling outputs may be produced [2, 3].

This chapter delves into the process of thermomechanical recycling, exploring its role in closed-loop textile recycling, its associated challenges and its role in Europe.

4.3.2 Thermomechanical Recycling of Textiles

Thermomechanical recycling, also known as polymer mechanical recycling, is a method involving the melting of thermoplastic synthetic polymeric material at high pressure and temperature, ultimately transforming it into pellets. In the context of textile thermomechanical recycling, this process utilizes synthetic post-industrial or post-consumer textile waste [4, 5].

For instance, textiles composed of pure synthetic materials such as polyamide 6, polyethylene terephthalate (PET), polyethylene, and polypropylene are amenable to thermomechanical recycling. However, the recycling of thermoplastic blend materials can prove to be highly challenging when the polymer materials exhibit different melting points or distinct flow properties [6–8].

It is crucial to emphasize that the textile waste, whether post-industrial or post-consumer, must consist of thermoplastic materials and must be free of any metallic elements in order to be suitable for this recycling method. For a complete textile-to-textile solution based on the thermomechanical process, a series of sequential steps must be followed, as illustrated in Figure 4.11. Thermomechanical recycling of textiles begins with the collection of post-industrial or post-consumer textile wastes. Once gathered, these wastes are sorted according to criteria such as material and color, and any non-target materials such as metals are meticulously removed. Sorting is followed by size reduction, which can be accomplished through methods, such as crushing, grinding, shredding, and pulling, to make the materials more manageable.

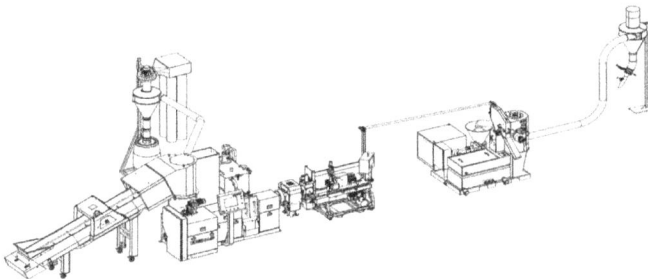

Figure 4.12 Thermomechanical process for transforming fabrics to pellets at CETI

Following size reduction, a thorough cleaning process is carried out to remove impurities and contaminants. With the aid of an automatic feeding conveyor, the fabric waste is cut, mixed, heated, dried, compacted and buffered in a pre-conditioning unit which is attached to a tangential extruder (see Figure 4.12). The extrusion process is filled continuously with hot, pre-compacted material. In the extruder screw the material is plasticized, homogenized and, if necessary, degassed in the venting zone. The melt is then cleaned in a fully automatic, multi-piston filter. Following this, the melt is conveyed to the respective pelletizer that turns polymer melt into high-grade, spherical, dry pellets.

For circular solutions, the pellets serve as the feedstock for both melt spinning and nonwoven applications. In melt spinning, a final step involves yarn post-texturization, which enhances the characteristics of the yarn, and ultimately, the creation of recycled fabrics through processes like knitting or weaving. This comprehensive textile recycling approach demonstrates the complexity of, and innovation involved in, achieving a textile-to-textile recycling system.

4.3.3 Current Status

Thermomechanical recycling is emerging as a promising avenue toward the attainment of a comprehensive textile-to-textile solution. However, it is imperative to consider the array of advantages and disadvantages associated with this process. On the positive side, thermomechanical recycling stands out as an economical, efficient, and well-established method. Contributing to its appeal are its ease of integration into existing operations that can produce nonwovens or filaments suitable for a wide range of textile applications.

On the negative side are certain limitations to be mindful of. The recycling of thermoplastic textile waste by the thermomechanical process causes polymer degradation, owing to the high temperatures and intense shear forces that are generated during extrusion. This degradation induces variations in key properties, including rheological and thermal properties, such as melting temperature and crystallization, and physical aspects, such as surface characteristics and color [8, 10, 11].

For example, polyester (PET) fabrics undergo a well-known reduction in intrinsic viscosity (IV) during thermomechanical recycling. The loss of PET IV is due to polymer hydrolysis and thermal decomposition. Therefore, PET IV must be enhanced via polymer blending or off-line polycondensation – which can significantly increase technology costs [10].

Furthermore, the recycling of post-industrial or post-consumer textile waste composed of a mixture of thermoplastic polymers presents significant challenges, due to disparities in their melt flows/viscosities, melting points, thermal characteristics, and recommended processing temperatures. To illustrate, consider post-industrial textile waste from the sock industry. This waste is typically composed of polyamide (PA) and thermoplastic polyurethane (TPU) threads. These two thermoplastic polymer materials differ substantially in their melting points and melt flow properties. The processing of such textile waste by the thermomechanical extrusion process requires a PA barrel temperature profile, which is significantly higher than that of TPU. Consequently, this elevated temperature adversely affects TPU, leading to severe extrusion problems, such as clogged filter screens, as shown in Figure 4.13, and backflows caused by a rise in pressure; these prevent the process from running continuously since breaks to replace the filter screens are needed. As a consequence, the resulting polymer blend exhibits subpar performance properties when it is used as feedstock for melt spinning applications [12].

Figure 4.13
Thermomechanical recycling:
clogged screens

Furthermore, the proper sorting and removal of metallic materials from textile waste are essential but often more expensive and complex for post-consumer textile waste compared with post-industrial waste. Post-consumer textile waste may have diverse material compositions and contamination, and these make thermomechanical recycling into textile fibers more challenging [6].

4.3.4 Ongoing Efforts

Different challenges are associated with thermomechanical textile recycling: textile sorting, polymer intrinsic viscosity (IV), filtration, and multicomponent materials. Various companies have been actively addressing these challenges and developing innovative methods to overcome them. These efforts have led to the creation of technologies that were originally designed for plastic thermomechanical recycling, but can also be applied to textile recycling [13–15].

For instance, companies have introduced filtration systems that allow polymer melt filtration with an automatic screen changer, ensuring a stable and continuous recycling process. Additionally, laser filtration systems have emerged as effective tools for decontamination in the recycling process. As mentioned above, the IV of the polymer often decreases after thermomechanical recycling, and the polymer must meet specific IV criteria for closed-loop recycling. Processes like multi-rotation systems, liquid or solid-state polymerization, and other polymer reactor techniques are utilized to boost the polymer's IV, making it suitable for textile applications. Online viscometers integrated with specific filtration systems enable the automatic segregation of polymer streams based on their viscosities. These categorized polymer streams can then be employed for various applications according to their IV levels [16–20].

Presently, companies are actively pursuing the development of a single-step thermome-chanical textile-to-textile recycling process, which combines all the necessary steps, in-cluding textile sorting, feedstock preparation, polymer filtration, IV enhancement and control, polymer pellet collection, and filament production. This integrated approach aims to achieve a continuous and efficient recycling process.

4.3.5 Thermomechanical Recycling in Europe

SCIRT is a H2020 project that brings together 18 frontrunners across Europe, including technical partners and research institutes, and clothing brands such as Decathlon, Petit Bateau, Bel & Bo, HNST and Xandres. The project aims to demonstrate a complete tex-tile-to-textile recycling system for discarded clothing or post-consumer textiles. Thanks to SCIRT, CETI (Centre Européen des Textiles Innovants) has had an opportunity to develop closed-loop recycling solutions for discarded textiles.

As shown in Figure 4.15, the fiber-to-fiber approach developed at CETI is a thermome-chanical recycling-based solution, which aims to provide circular solutions for post-production and postconsumer polyester textile fabric waste. The multi-step fiber-to-fi-ber approach includes the following processes:

- Conversion of fabric waste into pellets via thermomechanical recycling

- Control of polyester IV via polymer compounding

- Off-line rheometry

- A multicomponent melt spinning process

In the fiber industry, multicomponent fibers are known to enhance the material's per-formance, and are commonly used in the development of innovative solutions when various cross-sectional shapes are needed. To overcome the degradation and contam-ination, CETI is currently developing a fiber-to-fiber approach using a multicompo-nent yarn development configuration (see Figure 4.14).

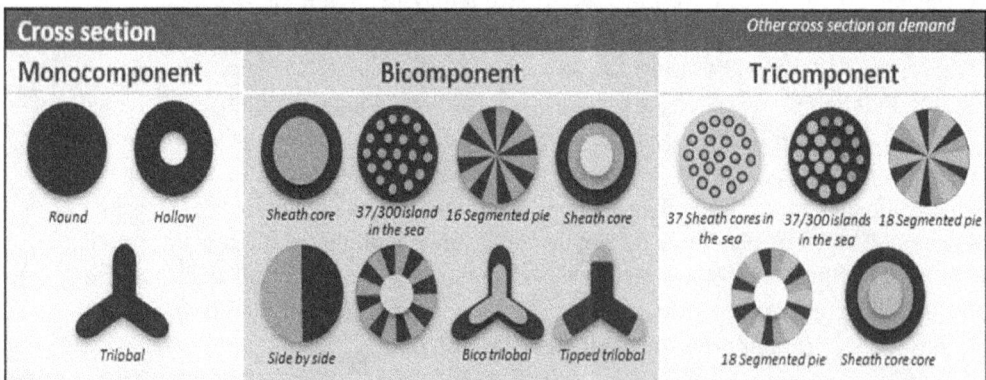

Figure 4.14 Possible fiber cross-sections in melt spinning at CETI

As the SCIRT project evolves, the project will also create new business opportunities by boosting textile value-chain activity and raise awareness of the environmental and social impact of buying clothes.

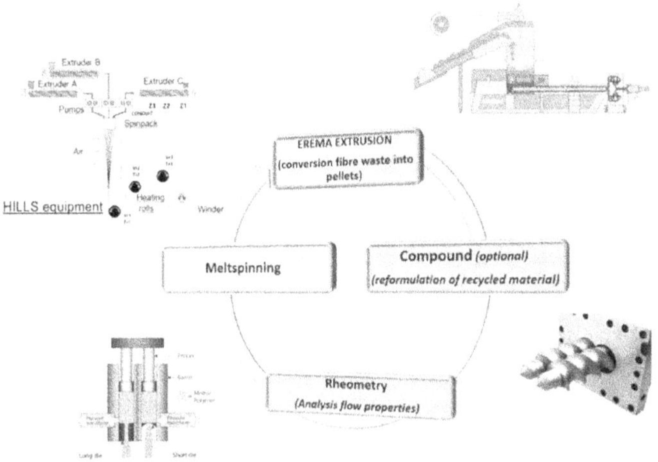

Figure 4.15 Textile circular solution: thermomechanical recycling in Europe

4.3.6 Conclusions

Thermomechanical recycling of textiles has emerged as a crucial process for addressing the escalating demand for textile products and the environmental challenges posed by the linear textile industry model. The textile industry has witnessed a substantial increase in production, driven by global demand, but this has also led to a significant disposal of underutilized textiles. The prevailing linear model, in which textiles are discarded rather than recycled, results in a substantial loss of resources, environmental inefficiencies, and potential increases in greenhouse gas emissions and microplastic pollution. Transitioning to a circular economy is imperative for maximizing material utilization and minimizing environmental impacts. Recycling is a pivotal component of this shift.

Textile recycling, however, is a complex challenge, due to the diversity of textile types and varying usage durations. The sorting of textiles plays a critical role and automated systems are being developed to improve the efficiency of this process. The introduction of technologies, such as NIR spectroscopy, holds out the promise of automating textile sorting on the basis of material type, color, and the detection of contaminants.

The thermomechanical recycling process is a key approach in the recycling of synthetic polymeric materials into pellets for reshaping into new fibers and other forms. This process involves several stages, including the collection and sorting of textile waste, size reduction, cleaning, densification, repelletization, yarn production, and fabric creation. It is important to note that the quality of the input feedstock, determined during the sorting phase, significantly influences the quality of the final recycled fibers.

The challenges of thermomechanical recycling include polymer degradation, especially in the case of post-consumer textile waste, as well as difficulties in processing blend materials with different melting characteristics. Nonetheless, efforts are being made to improve the efficiency of this recycling method and to maintain the quality of recycled materials.

The future of thermomechanical recycling looks promising as companies develop innovative technologies to overcome challenges. Filtration systems, including laser filtration, are enhancing decontamination and techniques to control and enhance polymer intrinsic viscosity are being employed. The vision is to create a single-step, efficient textile-to-textile recycling process that combines various stages for a continuous and rapid recycling system. This integrated approach aims to propel textile recycling further into the circular economy, minimizing waste, conserving resources, and reducing environmental impact.

Materials recycled through the thermomechanical process experience a loss of physical properties. In response to these challenges, CETI has taken a forward-looking approach by developing innovative material formulations for enhancing material properties. Moreover, CETI's multi-component fiber strategy has proven to be highly effective in ensuring the necessary mechanical performance at the yarn scale, ultimately facilitating successful closed-loop recycling. This section has underscored the significance of research and innovation in advancing sustainable practices within the textile industry.

References for Section 4.3

[1] Payne, A., Open- and closed-loop recycling of textile and apparel products. Handb Life Cycle Assess Text Cloth. 2015 Jan 1;103–23.

[2] Circle Economy, EigenDreads, Fashion for Good. Sorting for Circularity Europe. 2022;(September).

[3] Étude de caractérisation des flux entrants et sortants de centres de tri. 2023;

[4] Handbook of Recycling [Internet]. Handbook of Recycling. Elsevier; 2014 [cited 2023 Jun 10]. 497–501 p. Available from: *http://www.sciencedirect.com:5070/book/9780123964595/handbook-of-recycling?via=ihub=*

[5] Damayanti, D., Wulandari, L.A., Bagaskoro, A., Rianjanu, A., Wu, H., Possibility Routes for Textile Recycling Technology. 2021;

[6] Worrell, E., Reuter, M.A., Handbook of Recycling: State-of-the-art for Practitioners, Analysts, and Scientists. Handbook of Recycling: State-of-the-art for Practitioners, Analysts, and Scientists. Elsevier Inc.; 2014. 1–581 p.

[7] Bahlouli, N., Pessey, D., Raveyre, C., Guillet, J., Ahzi, S., Dahoun, A., et al., Recycling effects on the rheological and thermomechanical properties of polypropylene-based composites. Mater Des [Internet]. 2012;33:451–8. Available from: *http://dx.doi.org/10.1016/j.matdes.2011.04.049*

[8] Felgel-Farnholz, A., Thermo-Mechanical Recycling: Study on the Performance of Unstabilized and Stabilized Polyolefins During Extrusion. Master's Thesis, Johannes Kempler University Linz, 2021 Avaliable from: *https://epub.jku.at/obvulihs/content/titleinfo/6542505/full.pdf*

[9] Le, K., Textile Recycling Technologies, Colouring and Finishing Methods. The University of British Columbia, 2018 Avaliable from: *https://sustain.ubc.ca/sites/default/files/2018-25%20Textile%20Recycling%20Technologies%2C%20Colouring%20and%20Finishing%20Methods_Le.pdf*

[10] Colin, X., Tcharkhtchi, A., Thermal degradation of polymers during their mechanical recycling To cite this version : HAL Id : hal-02618344. 2020.

[11] Dorigato, A., Advanced Industrial and Engineering Polymer Research Recycling of polymer blends. Adv Ind Eng Polym Res [Internet]. 2021;4(2):53–69. Available from: *https://doi.org/10.1016/j.aiepr. 2021.02.005*

[12] Kunchimon, S. Z., Polyamide 6 and thermoplastic polyurethane recycled hybrid fibres via twin-screw melt extrusion. 2019;

[13] Extruder, melt filters, screen changer and sensors from Gneuss [Internet]. [cited 2023 Jun 17]. Available from: *https://www.gneuss.com/en/*

[14] EREMA plastic recycling systems [Internet]. [cited 2023 Jun 17]. Available from: *https://www.erema. com/*

[15] Polyester Renewal Technology | Eastman Circular Economy [Internet]. [cited 2023 Jun 17]. Available from: *https://www.eastman.com/Company/Circular-Economy/Solutions/Pages/Polyester-Renewal.aspx*

[16] Laser Filter | POWERFIL–EREMA Filter Systems [Internet]. [cited 2023 Jun 17]. Available from: *https:// www.powerfil.com/en/laserfilter/*

[17] Continuous Polycondensation Plant Solutions | Oerlikon [Internet]. [cited 2023 Jun 17]. Available from: *https://www.oerlikon.com/polymer-processing/en/solutions-technologies/polymer-processing/ continuous-polycondensation-plant-solutions/*

[18] Polyreaction PET–Gneuss [Internet]. [cited 2023 Jun 17]. Available from: *https://www.gneuss.com/ en/polymer-technologies/polyreaction/*

[19] Dynisco–ViscoSensor [Internet]. [cited 2023 Jun 17]. Available from: *http://www.dynisco.com/polymer-evaluation/online-rheological-testing/Online-Rheometers/ViscoSensor*

[20] Viscometer–The Gneuss Online Viscometer (VIS) [Internet]. [cited 2023 Jun 17]. Available from: *https://www.gneuss.com/en/polymer-technologies/extrusion/online-viscometer-vis/*

4.4 Thermomechanical Textile-to-Fiber Recycling of PET Waste

Robin Sujatta, Matthias Schmitz, BB Engineering GmbH, Remscheid, Germany

4.4.1 Feedstocks and the Challenges they Pose for Thermomechanical PET Recycling

Besides virgin PET, alternative feedstocks are gaining in importance in the production of textiles. Whereas presently only approximately 15% come from secondary sources, of which 99% are PET-bottle flakes, targets are much higher [1]. Thus, up to now thermomechanical recycling technologies have been the main processes for recycled polyester textiles. As food-grade flakes are becoming rare, due to exclusive usage in bottle-to-bottle recycling [1], the industry is seeking alternatives. In the context of *circularity*, textiles are the only reasonable recycling feedstock for new fibers. This can be divided into the two main categories of post-industrial and post-consumer waste, both of which pose their own individual challenges in thermomechanical recycling.

Post-industrial waste includes many stages from spinning (start-up lumps, freefall yarn, take-up waste, wound bobbins, staple cord) through texturizing to woven or knitted fabric cut-offs. This material is well sorted and there is good knowledge about the constituents, including the disadvantageous presence of spinning oil or non-PET components. The main challenge consists in handling the different shapes and bulk densities and to converting them into a form that the extruder can handle. Varying moisture contents, spinning oil content and IV levels require special treatment before extrusion.

On the other hand, post-consumer waste covers clothing and home textiles, which have been used by the end consumer. Here the biggest challenge is the unknown composition with regard to polymer mix and additives or coatings. Collection and sorting facilities are complex and not widely established. A positive subcategory is commercial waste, which includes textiles e.g., from the hospitality sector – well sorted and with a defined history. Return mechanisms can easily be implemented.

Both feedstocks have in common the fact that input viscosity levels (typically described as intrinsic viscosity IV) are too low, fluctuate too much and have a contaminant content that is too high for filament spinning. Consequently, the recycling process needs to homogenize and clean the polymer.

4.4.2 Process Requirements for Thermomechanical PET Recycling

The textile production process, especially for products with filament yarn components, is one of the most crucial as regards material properties. Within the recycling process and the filament or fiber production process for textile-to-fiber applications, a range of demanding requirements must be fulfilled. Primary requirements for the process are efficient decontamination, exact setting of the viscosity, material homogenization, overall low material stress and a small economic and ecological footprint (see Figure 4.16).

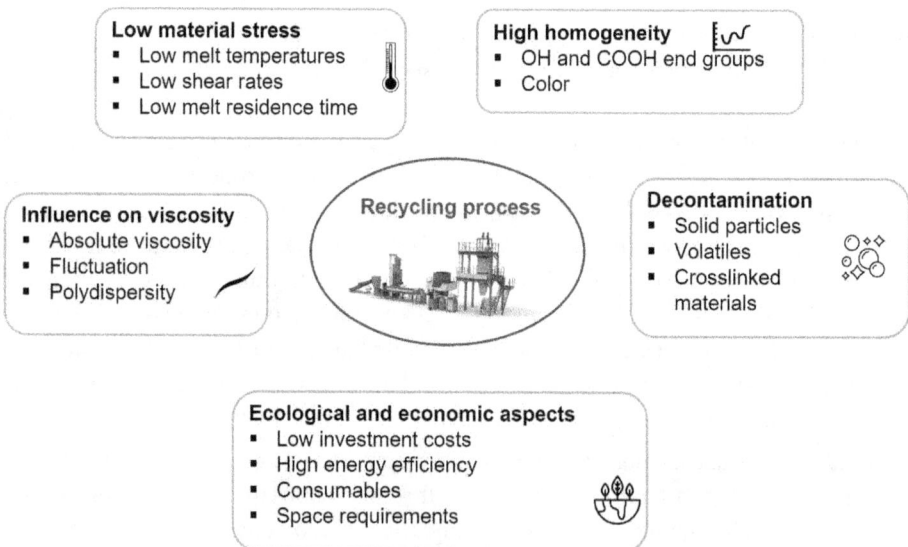

Figure 4.16 Process requirements for thermomechanical PET textile-to-fiber recycling [2]

Decontamination entails the removal of different components, such as solid particles (e.g., dust), viscous materials (e.g., crosslinked materials) and volatile components (e.g., spinning oils, additives). An effective decontamination is indispensable for guaranteeing high production performance in further processing steps and for meeting end-product specifications, such as color values and odor. For different types of textile applications, different intrinsic viscosity levels are needed from 0.6 dl/g to 1.2 dl/g. Due to this, an exact setting of the target viscosity as a function of the application is intended. Textile waste streams are mainly combining waste with different viscosities and therefore a stability of the output on a molecular basis (e.g., molecular weight, molecular weight fluctuation, polydispersity) is needed for further processing. Furthermore, the material homogeneity is crucial. In dependence of the application, properties like a low number of carboxylic end-groups or a minimized decoloring are significant.

To decrease the overall material stress during thermomechanical recycling, different requirements can be focused. A low melt temperature and decreased melt residence time are crucial for a high-quality output. Furthermore, low shear rates within the extrusion ultimately lead to improved material properties and less material degradation. Finally, economic, and ecological aspects such as energy efficiency, consumables, investment costs, labor and space requirements are criteria that need to be kept low to make recycling of textiles competitive [2].

4.4.3 Thermomechanical PET Textile-to-Fiber Recycling Process

This section focuses on the textile-to-fiber inline recycling process. In an inline process, the initial heat is used to plasticize the PET and directly form new filament yarns or fibers. Thus, pelletizing, a second drying and plasticizing within the yarn/fiber manufacturing process are not necessary [2].

The first process step of the textile-to-fiber recycling line consists of a conveyer belt with a metal detector and a cutting mill (see Figure 4.17). The fibrous materials are placed on the conveyor belt and fed into the cutting mill to reduce the size of the textile waste and to set the right bulk density for downstream process steps. After leaving the cutting mill, the material is conveyed into the extruder side-feed. Within this part of the system, the material gets compacted and is directly fed into the main degassing extrusion system. An optional hot air or desiccant drying system additionally removes the surface moisture so as to minimize hydrolytic degradation within the melt phase. Another possible feature at this point is the admixture of bottle flakes, masterbatch or other additives to the main textile waste stream. The shredding and direct feeding into the extruder has several advantages over other processes, like agglomeration. Key points are lower energy consumption, fewer material heating cycles and therefore less material degradation and decoloring [2].

The single-screw degassing extrusion system is particularly designed to minimize material stress during plasticizing. The main advantages are minimized shear rates and thus a decreased melt temperature and less material degradation. The main extrusion is followed by pre-filtration to remove all rough particles and protect further downstream components. Pre-filtration commonly uses a continuously working backflush screen changer. In this process step, rough particles are filtered from the polymer melt. Removing particles at an early stage of the process leads to less wear on the recycling equipment but also less material decoloring during recycling. The next process step is that of liquid state polycondensation (LSP). It removes volatile contamination and both increases and stabilizes the viscosity. Detailed working principles are described in Section 4.4.3.1 [2].

The subsequent crucial process step is large-area filtration in a candle-type filtration system with 20 µm filtration fineness. This system features a two-chamber arrangement, which guarantees non-stop operation. While one filter insert is in operation, the other insert is being cleaned. Note that high-tech filament yarn production for textiles can only be achieved through depth-filtration at low flow velocities and pressure differentials within the filtration medium [3]. The latest developments increase efficiency of fine filtration by allowing inline cleaning of the filter candles with super-heated steam. This method extends the filter's lifetime and reduces expenses. A deeper analysis of large-area filtration is provided in Section 4.4.3.2.

The recycling process is then finalized by directing the melt straight into a melt-spinning plant. Figure 4.17 shows a partially oriented yarn (POY) inline process. However, other PET fiber processes can be fed directly from the recycling plant (e.g., staple fiber, bulked continuous filament, fully drawn yarn, monofilament).

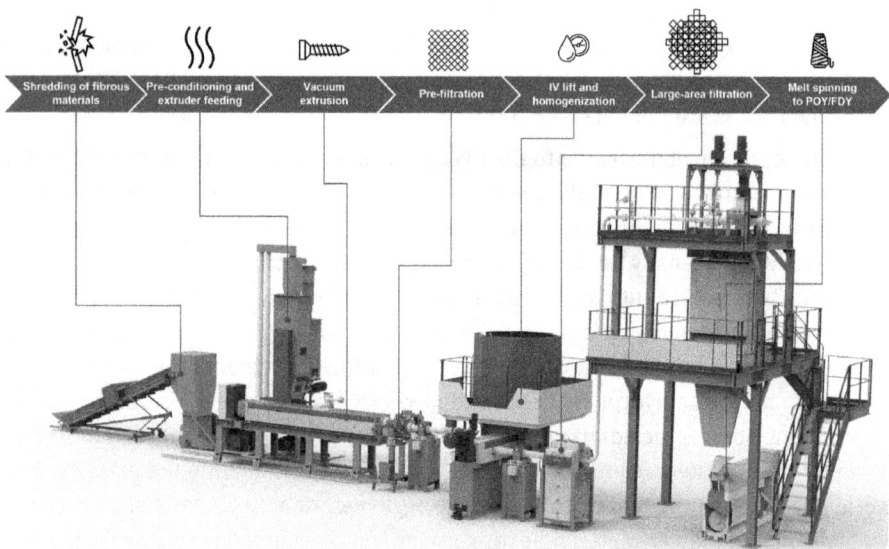

Figure 4.17 Thermomechanical PET textile-to-fiber recycling process [2]

4.4.3.1 Liquid State Polycondensation

Liquid state polycondensation is a reaction that occurs within the PET, whereby the polymer chains are extended and rearranged while being exposed to low vacuum levels within the melt phase. PET is often referred to as a living polymer. After the initial polycondensation to virgin polymer, both carboxylic and hydroxylic end-groups from the terephthalic acid and ethylene glycol are still reactive. During the LSP, water and glycol are removed from the polymer, giving rise to repolymerization through the formation of ester bonds. Compared with conventional solid state polycondensation (SSP) reactions used for bottle-to-bottle recycling, LSP has a significantly higher reaction speed, due to the higher temperatures. Conventional SSP processes are carried out at 200–240 °C for several hours. In comparison, an LSP process takes place at 270–295 °C for 5–60 min, a fact which ultimately leads to greater energy efficiency of up to 48%. Beyond that energy efficiency, high-molecular contaminants in textiles (e.g., spinning oils) cannot be removed sufficiently by means of SSP, due to its lower process temperature [2, 4–6].

Within the patented liquid state polycondensation, the polymer melt is led to a throughput-dependent number of candle filters, where it runs down within the vacuum chamber and eventually drops down into the feed zone of an extraction extruder or melt pump which builds up pressure for the following process steps [7]. The main parameters governing an efficient LSP process are:

- Melt temperature

- Vacuum level

- Melt residence time in the vacuum chamber

- Length of diffusion pathways for glycol

- Surface renewal rate

The higher the melt temperature within the LSP, the faster the polycondensation reaction takes place. An elevated temperature leads to higher diffusion speeds of the glycol within the PET polymer. However, if the reaction temperature comes too close to the degradation temperature of PET (> 300 °C) the opposite effects occur, leading to increased material degradation and loss of material properties. Another important process parameter of the LSP is the vacuum level. The reaction requires a vacuum of 1–5 mbar. The lower the vacuum level, the greater is the diffusion rate of the glycol within the polymer and the more efficient and faster is the degassing. Furthermore, increasing the overall residence time increases the viscosity uptake. The length of diffusion pathways and the surface renewal rate are mainly predefined by the operating principle of the LSP system. For example, BB Engineering's (BBE) LSP system *Visco⁺* creates a very thin polymer melt layer and consequently short diffusion pathways and a high surface renewal rate. Increasing both parameters accelerates the uptake in viscosity. Figure 4.18 shows the influence of the main LSP parameters [2].

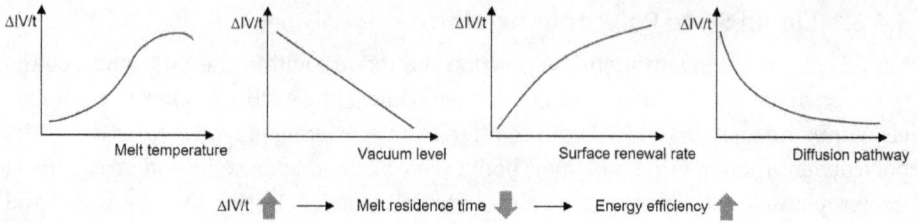

Figure 4.18 Main LSP parameters [2]

One of the most crucial parameters regarding material properties is the melt residence time. Since the melt residence time is directly linked to material degradation mechanisms, such as decoloring (CIElab, b-value), crosslinking, formation of acetaldehyde or an increase in the number of OH and COOH end-groups, a lower melt residence time is beneficial for further processing and product properties. Increasing the reaction speed allows the overall melt residence time to be lowered while still reaching the same target viscosity [2, 5].

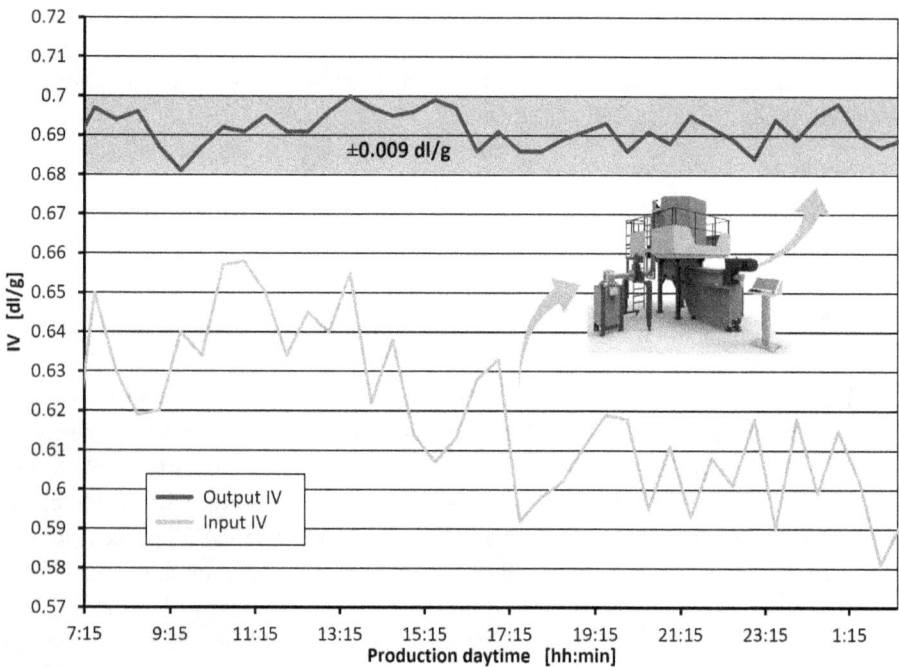

Figure 4.19 Typical LSP process [2]

A further important material property of recycled material is viscosity homogenization. When different feedstocks are mixed, they need to be harmonized to a homogeneous polymer. A minimum IV fluctuation of ± 0.01 dl/g is indispensable. The LSP reactor includes an online viscometer that regulates the vacuum level within the reactor between 1 and 5 mbar to influence the output viscosity. The shorter the residence time, the faster the system can react to compensate for fluctuations in input materials or residual moisture content. Figure 4.19 shows a typical production chart for a strongly fluctuating input material, namely short-term (5 min) fluctuations of up to 0.030 dl/g and long-term (3 h) fluctuations of up to 0.060 dl/g. The BBE *Visco*⁺ LSP system shown here is then able to reach the targeted IV of 0.695 dl/g with minimized fluctuations of 0.009 dl/g, a figure which meets in full the requirement for high-tech melt spinning applications. Thus, this LSP reactor is optimized for processes that impose high demands on viscosity requirements [2].

4.4.3.2 Melt Filtration for Textile Applications

An often-underestimated topic within recycling for textiles is that of melt filtration. A typical bottle-to-bottle recycling process requires a final filtration fineness of 40–60 μm for downstream injection molding and blow molding processes. By comparison, textile applications require a final filtration of down to 15 μm to guarantee an efficient spinning process. To realize a final filtration fineness, a multi-stage filtration process is needed. A pre-filtration unit is placed right after plasticizing of the PET in the extruder. Performing the filtration as soon as possible in this way minimizes decoloring because it prevents disadvantageous reactions between the polymer and the contaminant. Typical filtration finesses of the pre-filtration unit range from 30–100 μm. A typical pre-filtration is realized with a backflushing piston screen changer. Given the extensive literature available on screen changers, this filtration system will not be discussed further.

The second and very crucial filtration step is that of large-area filtration system. Such a system can cover filtration areas of up to 40 m². This system operates continuously without any need to stop production or risking a loss of material quality while operating. The filter is made of two filtration chambers (see Figure 4.21), each of which consists of material- and throughput-dependent number of pleated candle-type filter elements with a filtration fineness of around 20 μm [2, 8].

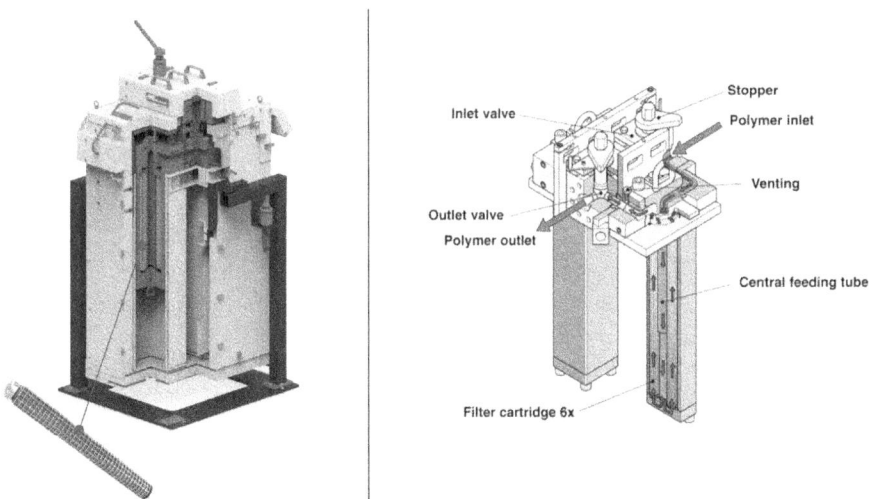

Figure 4.20 Large-area filtration system [8, 9]

As soon as the maximum differential pressure of the operating insert is reached, the second filtration chamber is filled with melt by switching the filter valves. Meanwhile, production continues to run uninterrupted. After filling of the clean filter insert with melt, production is fully switched to the second insert and cleaning of the used insert can begin. One of the key features of the filtration system is low flux rates.

$$\text{Flux rate} = \frac{\text{Throughput}}{\text{Filtration surface area}} \qquad (4.1)$$

Low flow velocities and differential pressures have the effect of not pushing particles bigger than the filtration fineness through the filtration medium. A further advantage is the removal of viscous (e.g., gels) and brittle materials (e.g., accumulated filler materials) from the polymer, which is not possible with conventional filtration methods at high flux rates. For example, the filter surface of a large-area filtration system is around 100 times bigger than common screen changers used for recycling applications. Thus, the flux rate of BBE's large-area filtration is around 100 kg/h/m^2 and that of a screen changer up to 10,000 kg/h/m^2. This ultimately leads to better filter pressure values (FPV) for the rPET and extends the spin pack lifetime for several days. FPV is defined as the differential pressure increase (Δp) while running a defined amount of polymer (m) through a defined filtration surface area (A) (see Equation 4.2, DIN EN ISO 23900-5). The lower the FPV is, the higher the quality of the material and the better the filtration system performs [3, 10].

$$FPV = \frac{\Delta p}{m \cdot A} \qquad (4.2)$$

The low stress on the spin packs with regard to contaminants not only extends pack lifetime but also boosts the spinning performance and the yarn properties. In particular, increasing the spin pack lifetime is a tremendous economic advantage, due to fewer cleaning cycles, which in turn leads to lower labor and material costs. Figure 4.21 shows the results of a filter pressure test performed according to ISO 23900-5 for PET bottle flakes with a 25 µm filtration fineness and virgin PET material. The recycled PET that has passed through a large-area filtration system not only far outperforms screen changers with 25 and 12 µm mesh size, but also has a filter pressure value comparable to that of common virgin PET materials for melt spinning applications [2].

The final filtration unit before the different kinds of fibers are formed is contained within the spin pack. For high-end filament yarn types (e.g., POY, FDY), *3LA-filters* are used (see Figure 4.22). These filters are specifically made for melt spinning of recycling materials. Because of their pleated mesh structure, the filtration surface is five to ten times higher than that of conventional spin pack sand filters and consequently results in an increased filter lifetime. These 3LA filters offer the advantage of a lower pressure increase and thus more uniform yarn properties [2, 11].

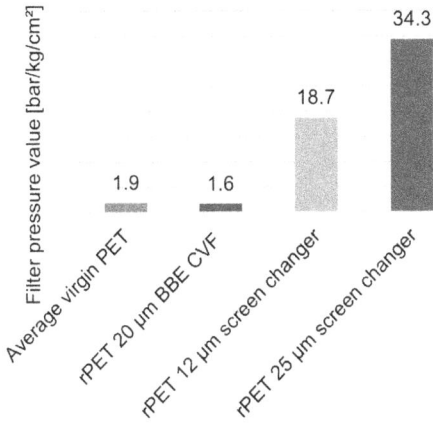

Figure 4.21
Filter pressure values with 25 μm for differently filtered rPET materials [2]

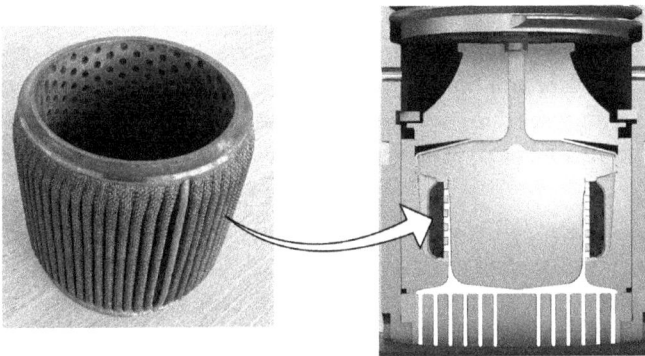

Figure 4.22 3LA filter used in POY/FDY spin pack filtration [2, 11]

4.4.3.3 Energy-Optimization Potential of the Inline LSP Process

The following section will compare the potential of an inline textile-to-fiber recycling process with LSP and a multi-step process with SSP.

State of the art in the last decade was a process in which firstly all different kinds of PET waste were agglomerated, plasticized with a degassing extruder, filtered once or twice with screen-changer technology and then pelletized. Without drying and LSP technology, the final IV was too low for use in spinning, so an additional solid state polycondensation step (SSP) was included. Finally, the chips were dried again, plasticized and spun into fibers and filaments. The fact that rPET was often only added in a certain percentage of 10–50% to virgin PET so that acceptable spinning performance could be achieved already shows the difficulties inherent in this recycling process. BBE's recycling solution addresses the main issues, which are inhomogeneity and insufficient filtration. Shrinking the process into a single-step inline process halved the energy consumption (see Figure 4.23) [2].

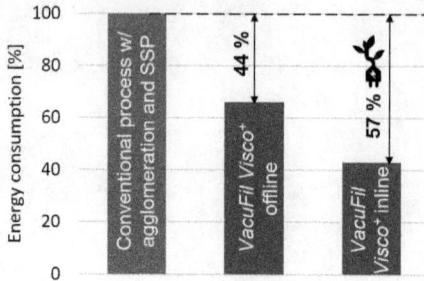

Figure 4.23

Energy consumption of different fiber-to-fiber recycling processes

In the first step, energy intensive agglomeration and SSP were replaced by a direct feeding of cut textiles into the extruder and LSP. Not only did this lead to a 44% reduction in energy consumption, but polymer degradation was also reduced through the use of the first heat of light drying and gentle extrusion. Degassing the polymer in the liquid phase leads to a much higher diffusion rate, resulting in easier extraction of volatile contaminants compared with SSP. The ability to measure the melt viscosity at any time during the process allows input fluctuations to easily be balanced by adapting the depth of vacuum. The result is a final IV fluctuation of less than ±0.01 dl/g (see Figure 4.19). As all components were designed for continuous operation, the intermediate steps of pelletizing, drying and extrusion can be skipped, which brings further energy savings of 13% [2].

In addition to low energy consumption, the presence of a large-area candle type filter improves the stability not only of spinning, but also of downstream processes, like texturizing and knitting. The increase in spinpack lifetime reduces waste generation, necessary manpower and cleaning costs by 70%. The latest innovations in inline filter cleaning further improve secondary processes, such as filter cleaning costs, and are paving the way toward a fully-automated process.

4.4.4 Opportunities for and Limitations of Thermomechanical PET Recycling – Application Examples

Thermomechanical recycling has great benefits for well-sorted PET waste feedstocks in terms of efficiency and simplicity.

On a production scale, many different textile feedstocks can already be validated for a 100% textile-to fiber-process, by producing filament yarn out of textile waste. They include not only post-industrial waste, such as circular knit cut-offs and mattress covers, but also post-consumer waste, such as jackets, sheets and shirts, from 100% PET. All yarn parameters, such as tenacity, elongation, dying uniformity and yarn-breakage-rate are within virgin specification. The tests even show better results than those of a discontinuous two-step process (offline) with additional pelletizing and extrusion. Compromises are necessary when it comes to color in post-consumer waste. The large range of colors (different shades of "white") and degree of degradation through washing or weathering lead to decoloring, which is reflected in an increasing b-value[1]. Different approaches aimed at reducing color change are extruder feeding under vacuum or nitrogen, dosing with liquid color, and adding TiO_2 masterbatch.

However, there are limiting factors acting on the process. Some ingredients tend to slow down and even block the polycondensation process. This effect can be seen in material containing co-polyester or coatings. Whereas dope-dyed yarn has only a minor influence on the process, bath dyeing or coating causes a rapid drop in IV to below 0.40 dl/g, as applied ingredients are unstable at melt temperatures above 280 °C.

4.4.5 Outlook

Without doubt, the biggest challenge in textile recycling is the availability of a consistent feedstock quality. Collection streams are still not established and sorting is still too complex for industrial scale. The majority of textiles are still a mixture of natural and synthetic fibers or of different polymers. Coatings and additives adversely affect the LSP process. Although *design for recycling* and a digital product passport, as suggested by the European Union [12], address these very topics, due to the long lifetime of textiles, it will take years or decades before the benefits are seen. In view of the fact that production of synthetic fibers occurs worldwide, textile waste is not a European but a global problem.

Technological innovation can lay the basis for circular textiles. Reducing production costs and improving resin quality can close the current price gap between virgin and recycled PET, and make recycling more attactive.

In post-production recycling, collaboration between producers can increase recyclable volumes and the feasibility of recycling systems.

References for Section 4.4

[1] Textile Exchange, Preferred Fiber & Materials Market Report, URL: *https://textileexchange.org/app/uploads/2022/10/Textile-Exchange_PFMR_2022.pdf* (2022)

[2] Sujatta, R., Schmitz, M., Kürten, P., *Ending fibrous polyester waste VacuFil Visco⁺ inline fiber-to-fiber recycling of PET waste*, Whitepaper (2023) 2, BB Engineering GmbH

[3] Alexander, J., *Recycling of PET* (2014) 10, Plastics in the Chemical Fiber Industry. Fiber Journal, pp. 30–33

[4] Mandal, S., Dey, A., (2019) Chapter 1–PET Chemistry. In: Thomas S (ed) Recycling of Polyethylene Terephthalate Bottles. William Andrew Publishing, New York

[5] Day, M., Wiles, D.M., (January 1972). "Photochemical degradation of poly(ethylene terephthalate). III. Determination of decomposition products and reaction mechanism". Journal of Applied Polymer Science. 16 (1): 203–215

---

[1] CieLab scale

[6]   Ma, Y., Agarwal, U.S., Sikkema, D.J., Lemstra, P.J., Solid-state polymerization of PET: influence of nitro-
       gen sweep and high vacuum, Polymer, Volume 44, Issue 15, 2003, Pages 4085-4096, ISSN 0032-3861

[7]   Alexander, J., Dickmeiß, F., Schmitz, M., Schäfer, K., Patent DE102017008320A1 (2017)

[8]   Schmitz, M., Presented at 62. Dornbirn Global Fiber Conference, Dornbirn, AT, September (2023)

[9]   BB Engineering GmbH, Polymer filters and white filter cleaning, URL: *https://bbeng.de/wp-content/
       uploads/2023/06/BBE_Filters_Broschuere_06-2023_web.pdf*

[10] DIN EN ISO 23900-5:2019-01, *Pigments and extenders–Methods of dispersion and assessment of disper-
       sibility in plastics–Part 5: Determination by filter pressure value test* (ISO 23900-5:2015); German version
       EN ISO 23900-5:2018

[11] BB Engineering GmbH, VarioFil R/R+ spinning line, URL: *https://bbeng.de/en/spinning-plants/variofil-r-r/*

[12] European Commission (2023), *https://environment.ec.europa.eu/strategy/textiles-strategy_en*

4.5 Chemical Recycling of Textile Waste – An Overview

Lukasz Debicki, Stefan Schonauer, Ricarda Wissel, Institut für Textiltechnik of RWTH
Aachen University, Germany

4.5.1 Chemical Recycling

The term chemical recycling is an umbrella term and is not uniformly defined. However, it can be used to summarize various processes for recycling textile residual materials. The main ones are solvent-based separation, thermochemical recycling depolymerization and biochemical recycling (see Figure 4.24).

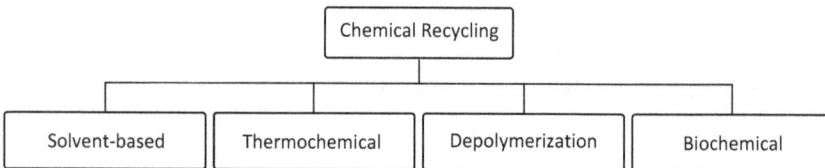

Figure 4.24 Overview of chemical recycling processes

Solvent-based separation can be further subdivided by solvent. In thermo-chemical
recycling, a distinction is made between pyrolysis and gasification. Depolymerization
can take place down to the oligomer or monomer level. Biochemical recycling is based
on dividing polymers with the help of enzymes. The individual processes are explained
in more detail below.

4.5.1.1 Thermochemical Recycling

Thermochemical recycling comprises two main processes: **pyrolysis** and **gasification**. In gasification, temperatures range from 700 to 1100 °C and the process occurs in
the absence of oxygen, leading to partial oxidation. Pyrolysis, in contrast, is carried out
at lower temperatures, between 350 and 700 °C, also without oxygen. Both processes

generate gas and oil, which can either be burned as fuel or purified and converted into chemical intermediates. Pyrolysis has a long history of use in plastics recycling, especially for fuel production. Gasification, meanwhile, has more recently been applied to the recycling of polyester-based carpets [1–2].

Pyrolysis is a process that breaks down organic waste in the absence of oxygen, turning it into three phases: gas, oil and solid residue. The syngas produced (a mix of CO and H_2) can be used as fuel or raw material to make alcohols and other hydrocarbons. Pyrolysis does not require pre-treatment and is ideal for contaminated waste. This is more efficient than biochemical methods, which often need chemicals, produce more waste and take longer to scale.

There are three main types of pyrolysis:

- Catalytic pyrolysis uses catalysts (such as metal oxides or activated carbon) to improve the process; this increases the yield and favors the production of specific products, such as olefins or aromatics

- Flash pyrolysis quickly heats waste to high temperatures (500–700 °C) and cools it rapidly, producing a liquid with a high energy value that is useful for making biofuels

- Slow pyrolysis heats biomass slowly (0.1 to 1 °C per second) to 400–500 °C over 5 to 30 minutes, creating solid carbon char and smaller amounts of gas and liquid

Each type of pyrolysis has its own benefits, which vary with the materials being processed and the desired outcome [3].

Thermal Gasification is a process that converts solid materials into a gas mixture called syngas, which mainly contains methane (CH_4), carbon dioxide (CO_2), carbon monoxide (CO), hydrogen (H_2), and tars. Syngas can be used as fuel in engines, for heating, electricity generation, and in chemical processes, such as the Fischer-Tropsch synthesis. The process of thermal gasification is exothermic and involves the transformation of carbon-based materials into syngas through a series of chemical reactions, including partial oxidation, the Boudouard reaction, the water gas reaction, the methanation reaction, complete oxidation, the oxidation of H_2, the water gas shift reaction and the steam reforming reaction. The composition of syngas varies with the residence time, pressure, temperature, feedstock type and the gasifier used. A long residence time helps increase hydrogen production and improve process efficiency. Different gasifying agents—air, steam, oxygen, or carbon dioxide—can be used in the process. Air is cheap, but dilutes syngas with nitrogen, reducing its energy content and efficiency. Oxygen avoids this problem, but increases costs. Steam is a more affordable alternative and it also boosts hydrogen content in syngas through reactions, such as steam-methane reforming. The use of CO_2 can also help control the hydrogen-to-carbon monoxide ratio and improve syngas quality, though it can raise costs [4].

Input/Output

Most waste can be processed through thermochemical recycling if it is made of organic or carbon-based materials, such as textiles. The process focuses on extracting carbon and hydrogen, and since many plastics contain high amounts of these elements, they are good materials for recycling. While some chemicals may affect the process, most contaminants do not cause major issues.

The various fractions emerging from the process can be viscous liquids (also called oil or tar fraction), condensable and non-condensable/permanent gases (syngas) and inorganic residue, such as carbon soot, metals and minerals (also called ash). The specific output depends on the input waste (type and composition). The obtained fractions can serve as feedstock for the chemical industry, and also as fuel [1–2].

Significance for Textiles

Gasification has long been used to convert mainly coal and biomass into energy. Today, companies are developing thermochemical methods to recycle plastics and textile waste. While gasification is a well-established technology, recent developments are focused on producing raw materials for the chemical industry, rather than just energy or fuel. Although only a few gasification processes have been tested, some are already operating as industrial plants. Gasification is especially useful for textile waste that cannot be recycled mechanically or thermomechanically and for handling small amounts of textile fractions. Given the need for effective recycling of difficult waste and the production of high-quality materials, gasification could become an important complementary technology. Its main advantage is the ability to handle mixed and contaminated waste, offering more feedstock flexibility than other recycling methods [1–2].

4.5.1.2 Depolymerization

Depolymerization is the breaking down of polymers into monomers or oligomers and reassembling them into virgin polymers (repolymerization). There are many different types of depolymerization. The particular type to be used depends on the polymer. This chapter focuses on the two input polymers polyethylene terephthalate and polyamide 6, which are the most important for the textile industry.

Polyethylene Terephthalate (PET) is a polyester containing ester groups that can be broken down by various reagents, including water, alcohols, glycols, acids, and amines. As a result, chemical recycling processes for PET are typically classified into the following types: hydrolysis (using water), glycolysis (using glycols), methanolysis (using alcohols) and other methods, such as aminolysis and ammonolysis (using amines) (see Figure 4.25) [5].

The products obtained from chemical recycling of PET depend on the reagent used. The various process options for recycling PET waste can be grouped into the following categories [5]:

- Conversion into oligomers (via glycolysis or solvolysis)
- Repurposing of glycolyzed waste for value-added products

- Conversion into specialty intermediates for use in plastics and coatings

- Regeneration of base monomers (methanolysis yields dimethyl terephthalate (DMT), and hydrolysis yields pure terephthalic acid (PTA) and ethylene glycol (EG))

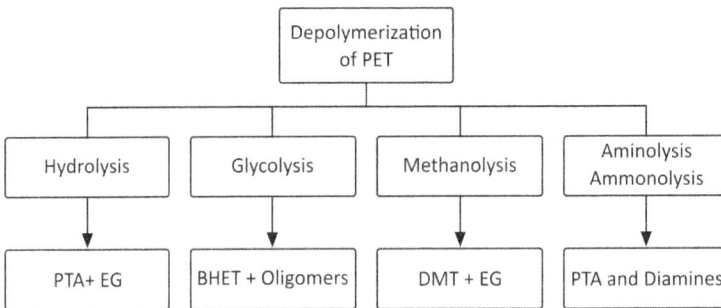

```
                    ┌──────────────────┐
                    │  Depolymerization │
                    │      of PET       │
                    └──────────────────┘
        ┌──────────────┬──────────────┬──────────────┐
 ┌────────────┐ ┌────────────┐ ┌────────────┐ ┌────────────┐
 │ Hydrolysis │ │ Glycolysis │ │ Methanolysis│ │ Aminolysis │
 │            │ │            │ │            │ │ Ammonolysis│
 └────────────┘ └────────────┘ └────────────┘ └────────────┘
 ┌────────────┐ ┌────────────────┐ ┌────────────┐ ┌────────────────┐
 │  PTA+ EG   │ │ BHET + Oligomers│ │  DMT + EG  │ │ PTA and Diamines│
 └────────────┘ └────────────────┘ └────────────┘ └────────────────┘
```

Figure 4.25 Processes for the depolymerization of PET and output materials. PTA: purified terephthalic acid, MEG: monoethylene glycol, DMT: dimethyl terephthalate, BHET: bishydroxylethylene terephthalate

Polyamide 6 (PA6)

Similar to the processes of PET recycling, PA6 can be depolymerized by hydrolysis, ammonolysis and glycolysis [6]. In practice, PA6 is commonly depolymerized by hydrolysis methods, such as high-pressure steam, acid hydrolysis with super-heated steam, or glycolysis [7].

Hydrolysis uses water molecules to break the carboxyl amide groups of PA6. This can be sped up by using high temperatures and pressures (thermal hydrolysis), acid catalysts (acid hydrolysis) or alkaline reagents (alkaline hydrolysis). Ammonolysis breaks PA6 polymer chains through reaction with ammonia at high temperatures (280–330 °C). Glycolysis uses the nucleophilic properties of hydroxyl groups and the high boiling point of glycols to break the amide bonds of PA6 and create monomers. However, due to low yields and byproducts, glycolysis is not used industrially [8].

Input/Output

Any PA6 or PET textiles can, in theory, serve as *input* material for depolymerization processes. PET is mainly recycled from food packaging and industrial waste, while textile recycling is still undergoing development. PA6 is primarily recycled from post-consumer materials, such as carpets, fishing nets and industrial waste, including oligomers and plastic waste from polymer industries. To an extent depending on the process, small amounts of other materials, such as dyes or coatings, are allowed. However, for cost reasons, most technology providers require at least 80–90% PET or PA6 content.

The traditional products derived from PET are pure terephthalic acid and monoethylene glycol, but the final *output* depends on the reagent used. The main product options include:

- Regeneration of base monomers by methanolysis for dimethyl terephthalate and hydrolysis or enzymatic methods for producing pure terephthalic acid and ethylene glycol

- Conversion into oligomers and bis(hydroxyethyl) ester of pure terephthalic acid (BHET) by glycolysis

The main output from PA6 recycling is ε-caprolactam, a ring-shaped monomer with six carbon atoms [7].

In general, *post-treatment* of the extracted materials is necessary. This usually consists of one or more of the following steps [7]:

- Purification (separation of contaminants; removal of colorants)

- Separation (distillation of caprolactam or removal of solvent via evaporation)

- Crystallization of BHET/PTA

- Drying

The monomers and oligomers obtained can then be repolymerized to produce high-purity, virgin-grade PET or PA6. An overview of the necessary process steps for recycling PET and PA by means of depolymerization can be seen in Figure 4.26.

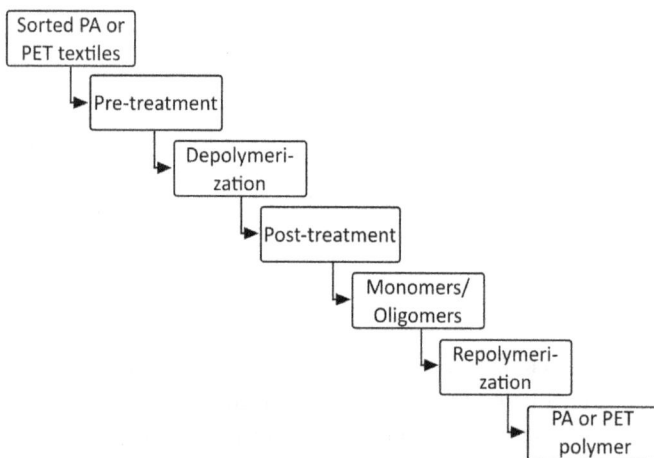

Figure 4.26 Process steps for recycling PET and PA by means of depolymerization [7]

Significance for Textiles

- Chemical recycling of PA6 textiles through depolymerization is a well-established technology that has reached TRL 9 and has been in use for over a decade [7]. For

PET textiles, the TRL levels vary from 4 to 9, with the level depending on the technology used. The major challenges lie in the economic efficiency of the processes and the handling of foreign materials. There are many processes at low TRL levels that are very promising. However, further development is needed to achieve the necessary scaling. The relevance of depolymerization for textiles is very high, as the recovery of raw materials in virgin quality enables multiple recycling.

4.5.1.3 Solvent-Based Recycling

Dissolution and reprecipitation technology refers to a recycling approach that aims to dissolve materials in specific solvents or solvent mixtures to separate their components. The basic principle of this technology is that the solvent acts selectively on the polymer to be separated, while other materials remain undissolved. This process is purely physical. The polymer is dissolved without change and is later precipitated again, with its original chemical structure retained. This distinguishes it from the more familiar, but erroneously similar term chemical recycling, which breaks down the polymer chain. The key to successful dissolution obviously lies in using the right solvent or the right solvent mixture of sufficient selectivity for the target polymer and only the target polymer. The solubility can be estimated by calculating the Hansen solubility parameters. These can considerably accelerate the screening of many solvent mixtures before they are tested experimentally [9–10].

In the dissolution process, the waste is pre-sorted to ensure that it contains as much of the target polymer as possible. If foreign polymers are also expected to be soluble in the solvent, they can be reduced or completely separated. The sorted waste is then placed in a suitable solvent and the target polymer is dissolved. Insoluble residues (polymers, organic and inorganic foreign substances) remain as solids and are filtered off. The remaining solvent then contains only the target polymer, which can be recovered as a solid, e. g. by evaporating the solvent or adding a non-solvent/precipitant.

Advantages of the process are:

- Material separation and thus higher output quality
- No depolymerization and repolymerization needed
- Virgin-like polymers
- Polymer in solution is very accessible and modifiable, e. g. dye removal or chain extension

Disadvantages of the process are:

- Not always possible with non-hazardous solvents
- Solvent recovery can be energy expensive

In principle, the process is suitable for all readily soluble polymers. However, it is currently mainly used for urethanes (elastanes), vinyl polymers, PET, PAs and cellulose.

Urethanes/Elastanes

One of the most common polymers in textiles, not by quantity but by distribution, is elastane. It is usually used in 5–20% to increase the wearing comfort or to improve the body fit. This makes it a common by-polymer and a problematic contaminant for recycling. However, elastane is soluble in various solvents, such as dimethyl acetamide, dimethyl formamide (DMF), dimethyl sulfoxide (DMSO) and tetrahydrofuran, and can thus be removed from the textile by dissolution. One of the best-known publications on this is by Boschmeier et al., who use DMSO, which is the only non-toxic member of the solvents named above, to remove elastane from PET/EL or PA/EL blends [11]. When removing elastane as a secondary component, the amount of solvent and the removal of the solvent from the main component is particularly critical. So far, this process has only been applied in laboratory-scale research.

Vinylic Polymers / Polyacrylonitrile (PAN)

Since polyacrylonitrile is a wet-spun polymer, dissolution of PAN waste is a particularly suitable recycling method. This means that PAN can be extracted from a textile mixture and processed directly as a spinning solution. There is no need for precipitation and drying. The common solvents for wet spinning of PAN, DMSO, dimethyl acetamide and DMF are suitable, but DMF is unfavorable as it causes yellowing. In addition, the cascade of washing baths in fiber production offers very interesting potential for treating the still gel-like fibers. For PAN, the process is mainly used in industrial settings to handle production waste. Research has also applied it to end-of-use waste and to remove dyes from the solution [12–14]. However, the packaging industry, which is larger in terms of volume, is more advanced here. Other vinyl polymers, such as polystyrene, polyvinyl chloride and acrylic butyl styrene, offer interesting concepts for PAN recycling. There are already established methods for recycling these polymers, including the removal of fire retardants and chain extension in solution [15–19].

Polyethylene Terephthalate (PET)

As it is a polycondensate of terephthalic acid and ethylene glycol, PET can be broken down into these monomers relatively easily. Common processes in addition to mechanical and thermomechanical recycling are therefore chemical depolymerization and enzymatic depolymerization. Of these, however, only depolymerization yields virgin-grade PET. Dissolution is interesting for PET, because it can recover virgin-grade PET without the need for depolymerization and repolymerization. The choice of solvent also depends on the waste stream, the cost of the solvent, the environment and safety. Suitable solvents for PET are dimethyl sulfoxide (DMSO), N-methylpyrrolidone (NMP) and sulfolane [20–21].

Polyamide (PA)

The most common polyamides in textiles are PA6 and PA6.6. PA6 is produced from a ring-shaped monomer, caprolactam, while PA6.6 is copolymerized from two monomers (adipic acid and hexamethylenediamine). The recycling processes for these materials therefore also differ. Mechanical and thermomechanical recycling are possible for both materials. PA6 can, in addition to chemical methods, also be thermally depolymerized back to the monomer. For PA6.6, chemical processes yield the monomer salts. Thanks to the amide bonds, enzymatic and biological recycling are also possible. These are the most common processes, but dissolution is also possible as a separation technique for PAs. Possible solvents here are formic acid, phenols, trifluoroethanol, α-cyanohydrin, hexamethyl-phosphoric triamide, phenol:tetrachloroethane, DMSO and $CaCl_2$-EtOH-H_2O mixtures [22–23].

Cellulose/Cotton

Cotton is composed of over 90% cellulose, a high-quality natural polymer. This makes it an ideal material for recycling, as processes for cellulose regenerated fibers can be used to produce high-quality fibers from cotton. The REFIBRA™ Technology by Lenzing is probably the highest TRL application for fiber recycling. In this process, cotton from production waste replaces approx. 30% of wood chips in a lyocell spinning process. This uses the ionic liquid N-methylmorpholine-N-oxide (NMMO) which can dissolve cellulose [24].

Significance for Textiles

Dissolution is an interesting process for separating polymers from mixed materials and reprocessing them into high-quality granulates and fibers. In terms of process complexity, it lies between mechanical, thermomechanical processes and chemical/biochemical recycling. In addition, the process can deliver virgin-grade polymers, which is otherwise only possible with chemical/biochemical methods. However, there are still a number of unanswered questions regarding which solvent is most suitable for which waste stream, the costs, environmental impact and occupational safety with the respective solvents as well as their recovery.

4.5.1.4 Biochemical Process

In biochemical recycling processes, the targeted substrate is broken down by enzymes and the degraded components are then utilized as raw materials for new products. Enzymes act as biological catalysts for chemical reactions while not being consumed or altered during the course of the reaction they accelerate. Before the catalytic function can be carried out, enzymes bind very selectively to their respective substrates. The region of the enzyme that binds the substrates is called the active site and represents a relatively small area of the overall molecular structure. The chemical reaction, for example a depolymerization, is triggered by the enzyme once the substrate enters the active site. Afterwards, the resulting products are released and the active site is free again to interact with a new substrate [25]. A schematic illustration of the interaction between enzyme and substrate is shown in Figure 4.27.

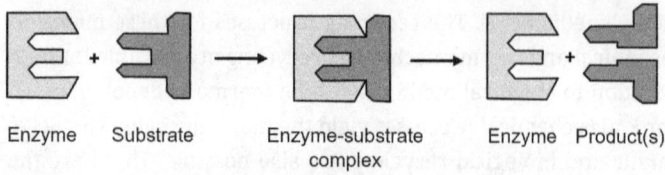

Enzyme   Substrate   Enzyme-substrate   Enzyme   Product(s)
                               complex

Figure 4.27 Schematic illustration of the interaction between enzyme and substrate (adapted from [25])

The method of enzymatic depolymerization can be applied to various types of textile fibers, including both natural and synthetic fibers. Due to their high selectivity, the choice of enzyme depends on the type of fiber material to be treated. This selectivity in breaking specific chemical bonds also qualifies enzymatic processes to be suitable for separating and recycling blended textiles which consist of at least two different fiber materials. While the selective enzymes break down the target fiber, the other fiber material remains intact during the process. For this reason, biochemical recycling is also seen as a promising approach for the treatment of polycotton, one of the most common blends in the textile industry containing cotton and PET fibers [26–27].

The process chain for enzymatic recycling starts with a pre-treatment of the textile substrate to make it more accessible to the enzymes and thus increase the depolymerization rate. As smaller particles represent a higher specific surface area, the probability that the active sites of the enzymes bind to accessible parts of the substrate is increased. Therefore, the substrate should be comminuted before enzymatic treatment, e. g. by grinding, cutting or shredding. In addition, it has been shown that the enzymatic degradation of amorphous substrates takes place at much higher reaction rates compared to crystalline substrates, so that a crystallinity-reducing pre-treatment can further facilitate subsequent degradation. The pre-treated substrate is then exposed to an enzyme-containing solution in a bioreactor. Since enzymes require specific conditions to be active, such as suitable temperatures and pH levels, the reaction conditions in the bioreactor must be carefully controlled. The enzymatic cleavage of bonds facilitates depolymerization, breaking down long polymer chains into smaller units. The resulting fragments are extracted and purified to remove remaining impurities, including residual dyes and other fibers [25, 28–29].

A notable example is the enzymatic hydrolysis of PET substrates for biochemical recycling. In contrast to other PET applications, such as films, PET fibers and textiles exhibit a high crystallinity of at least 30%. Given that PET becomes nearly resistant to enzymatic degradation when its crystallinity exceeds 20%, pre-treatment to prepare a more amorphous substrate is crucial. This can be achieved through melt quenching, in which high-crystalline PET is re-melted in an extruder and subsequently subjected to rapid cooling in a water bath. Due to the fast cooling, the time available for the polymer chains to organize into a crystalline structure is limited, so that they remain mostly in the amorphous state. Various hydrolase-type enzymes have demonstrated their ability to degrade

PET by cleaving ester bonds, e. g. PETases [30]. The end-product of the enzymatic degradation is the hydrolysate containing the monomers terephthalic acid (TPA) and ethylene glycol (EG). After separation and purification, these monomers can be repolymerized into recycled PET, which is suitable for the same applications as virgin PET [26]. Figure 4.28 shows the closed-loop process chain for biochemical recycling of PET textiles.

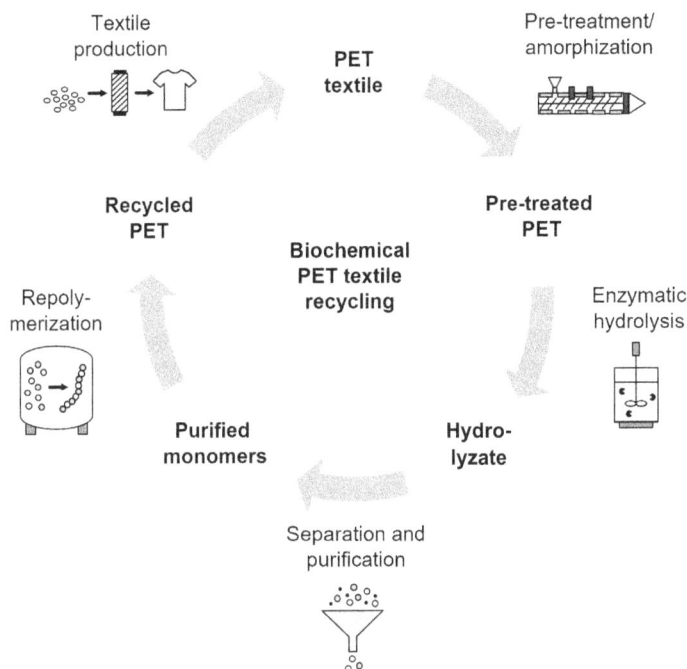

Figure 4.28 Closed-loop process chain for biochemical recycling of PET textiles

Different research approaches focus on the enzymatic hydrolysis of cotton fibers from polycotton textiles. The pre-treatment involves shredding the textile waste into smaller fragments, followed by a chemical treatment with a 20% sodium hydroxide (NaOH) solution to increase the efficacy of the subsequent hydrolysis process. Cellulase enzymes have the ability to degrade the polymer chains of cellulosic materials like cotton, leading to the production of the monomer glucose. Glucose is a valuable platform chemical for conversion to various products, such as ethanol. Starting from polycotton textile waste, the cellulosic fiber material can be removed enzymatically, while PET remains intact and is separated by filtration. The recovered PET can then be processed into a new textile through repelletization, melt spinning and textile surface production, e. g. weaving [28].

In summary, biochemical recycling methods have the potential to address challenges related to textile waste, especially blended fiber materials. However, these technologies are not yet implemented on a commercial scale. Intensive research is being conducted in this area, including pre-treatment of textile waste and enzyme engineering, and initial demonstration facilities for biochemical recycling of PET are being established by companies. In contrast to other chemical recycling processes, which often require high temperatures and pressures, enzymatic hydrolysis can operate under relatively mild reaction conditions and achieve equally high output material quality [27, 30].

4.5.1.5 Conclusion

Chemical recycling covers a range of processes. These can be grouped as follows: solvent-based, thermochemical, depolymerization and biochemical. Chemical recycling is relevant for a wide range of fiber types. Both man-made fibers, such as PET and PA fibers and natural fibers, such as cotton, can be chemically recycled. At present, high demands are still placed on the purity of the input materials. Elastane in particular represents a major disruptive factor for many processes, both in process control and in the resulting waste streams. The high purity requirements are one reason why chemical recycling is not yet economically viable. In the coming years, recycling of cotton or cellulosic fibers may achieve feedstock acceptance at purity levels of 70%, whereas recycling technologies for polyester and polyamide could demand purity levels exceeding 80%.

References for Section 4.5

[1]   Roos, C. J., (2010), Clean Heat and Power Using Biomass Gasification for Industrial and Agricultural Projects, U.S. Department of Energy *https://www.energy.wsu.edu/documents/biomassgasification_2010.pdf*

[2]   Pohjakallio, M., Vuorinen, T., Oasmaa, A. "Chapter 13–Chemical routes for recycling—dissolving, catalytic, and thermochemical technologies." In Plastic Waste and Recycling, edited by revor M. Letcher, 359–384. (2020) Academic Press. doi: *https://doi.org/10.1016/B978-0-12-817880-5.00013-X.*

[3]   Hussain, M., Ashraf, M., Kaleem Ullah, H., Akram, S. (2023). Recycling in Textiles. In: Batool, S.R., Ahmad, S., Nawab, Y., Hussain, M. (eds) Circularity in Textiles. Textile Science and Clothing Technology. P. 189 Springer, Cham. *https://doi.org/10.1007/978-3-031-49479-6_7*

[4]   Santos, S. M., Assis, A. C., Gomes, L., Nobre, C., Brito, P. (2022, December). Waste gasification technologies: a brief overview. In Waste (Vol. 1, No. 1, pp. 140–165). MDPI. *https://doi.org/10.3390/waste1010011*

[5]   Achilias, D., Andriotis, L., Koutsidis, I., Louka, D., Nianias, N., Siafaka, P., Tsagkalias, I. S., Tsintzou, G., 2012. "Recent Advances in the Chemical Recycling of Polymers (PP, PS, LDPE, HDPE, PVC, PC, Nylon, PMMA." In Material Recycling–Trends and Perspectives, edited by Dimitris S. Achilias. IntechOpen. doi: 10.5772/33457.

[6]   Minor, A., Goldhahn, R., Rihko-Struckmann, L., Sundmacher, K., (2023). Chemical Recycling Processes of Nylon 6 to Caprolactam: Review and Techno-Economic Assessment, Chemical Engineering Journal, Volume 474, 145333, ISSN 1385-8947, *https://doi.org/10.1016/j.cej.2023.145333*

[7]   Duhoux, T., Maes, E., Hirschnitz-Garbers, M., Peeters, K., Asscherickx, L., Christis, M., Stubbe, B., Colignon, P., Hinzmann, M., Sacheva, A., (2021). Study on the technical, regulatory, economic and environmental effectiveness of textile fibres recycling Final Report. 10.2873/828412.

[8]   Tonsi, G., Maesani, C., Alini, S., Ortenzi, M. A., Pirola, C., 2023, Nylon Recycling Processes: a Brief Overview, Chemical Engineering Transactions, 100, 727–732. *https://doi.org/10.3303/CET23100122*

[9] Yi-Bo Zhao, Xu-Dong Lv, Hong-Gang Ni,Solvent-based separation and recycling of waste plastics: A review, Chemosphere, Volume 209, 2018, Pages 707–720, ISSN 0045-6535, *https://doi.org/10.1016/j. chemosphere.2018.06.095.*

[10] Miller-Chou, B. A., Koenig, J. L., A review of polymer dissolution, Progress in Polymer Science, Volume 28, Issue 8, 2003, Pages 1223-1270, ISSN 0079-6700, *https://doi.org/10.1016/S0079-6700(03)00045-5.*

[11] [11Boschmeier, E., Archodoulaki, V.-M., Schwaighofer, A., Lendl, B., Ipsmiller, W., Bartl, A., New separation process for elastane from polyester/elastane and polyamide/elastane textile waste, Resources, Conservation and Recycling, Volume 198, 2023, 107215, ISSN 0921-3449, *https://doi. org/10.1016/j.resconrec.2023.107215.*

[12] *https://acrycycle.com/en/index* (2023)

[13] *https://www.birlacril.com/* (2023)

[14] Schonauer, S. (2020) industrial RePAN, *https://www.ita.rwth-aachen.de/global/show_document.asp?id= aaaaaaaaatskquj*

[15] Polystyrene Recycling: Polystyvert has developed a unique dissolution technology for recycling polystyrene. (2023) *https://polystyvert.com/en/technology/*

[16] [16García, M. T., Gracia, I., Duque, G., de Lucas, A., Rodríguez, J. F., Study of the solubility and stability of polystyrene wastes in a dissolution recycling process, Waste Management, Volume 29, Issue 6, 2009, Pages 1814-1818, ISSN 0956-053X, *https://doi.org/10.1016/j.wasman.2009.01.001.*

[17] Catalisti, DISSOLV: Pioneering circular PVC with INEOS Inovyn *https://www.catalisti.be/en/news/dissolv-pioneering-circular-pvc-ineos-inovyn*

[18] VinyLoop: Positive Ökobilanz für PVC-Recycling, *https://www.kunststoffe.de/a/news/vinyloop-positive-oekobilanz-fuer-pvc-re-280768*

[19] Balan, A., Bouquet, G., Eekman, E., Lakeman. P., (2022) Recycling method for elastomer toughened thermoplastic polymers, Patent: WO2022144158A1 or EP4267664B1

[20] Muringayil Joseph, T., Azat, S., Ahmadi, Z., Moini Jazani, O., Esmaeili, A., Kianfar, E., Haponiuk, J., Thomas, S., Polyethylene terephthalate (PET) recycling: A review, Case Studies in Chemical and Environmental Engineering, Volume 9, 2024, 100673, ISSN 2666-0164, *https://doi.org/10.1016/j. cscee.2024.100673.*

[21] Poulakis, J. G., Papaspyrides, C. D., Dissolution/reprecipitation: A model process for PET bottle recycling, (2021) Journal of Applied Polymer Science: Volume 81, Issue 1, Pages 91–95, *https:// onlinelibrary.wiley.com/doi/10.1002/app.1417*

[22] Goldhahn, R., Minor, A.-J., Rihko-Struckmann, L., Ohl S.-W., Pfeiffer, P., Ohl, C.-D., et al., Recycling of Bulk Polyamide 6 by Dissolution- Precipitation in CaCl2-EtOH-H2O Mixtures. ChemRxiv. 2024; doi:10.26434/chemrxiv-2024-8z1pm

[23] Hirschberg, V., Rodrigue, D., Recycling of polyamides: Processes and conditions (2023), Journal of Polymer Science, Volume 61, Issue 17, Pages 1937–1958, https://doi.org/10.1002/pol.20230154

[24] Tencel, REFIBRA™ Technologie (2024), *https://www.tencel.com/b2b/de/technologies/refibra-technology*

[25] Khan, M. Y., Khan, F., Principles of enzyme technology.–Delhi: PHI Learning Private Limited, 2015.

[26] European Commission: Study on the technical, regulatory, economic and environmental effective-ness of textile fibres recycling. Final Report. (2021) *https://www.ecologic.eu/sites/default/files/publication/ 2022/50030-study-textile-recycling-web.pdf*; Zugriff: August 2023

[27] McKinsey & Company: Scaling textile recycling in Europe – turning waste into value. (2022)

[28] Piribauer, B., Bartl, A., Ipsmiller, W., Enzymatic textile recycling – best practices and outlook. Waste Management & Research, 2021, *https://doi.org/10.1177/0734242X211029167*

[29] Chang, A.C., Patel, A., Perry, S., Soong, Y.V., Ayafor, C., Wong, H.-W., Xie, D., Sobkowicz, M.J., Understanding consequences and tradeoffs of melt processing as a pretreatment for enzymatic depolymerization of poly(ethylene terephthalate). Macromolecular Rapid Communications, 2022, *https://doi.org/10.1002/marc.202100929*

[30] Cheng, Y., Cheng, Y., Zhou, S., Ruzha, Y., Yang, Y., Closed-loop recycling of PET fabric and bottle waste by tandem pre-amorphization and enzymatic hydrolysis. Resources, Conservation & Recycling, 2024, *https://doi.org/10.1016/j.resconrec.2024.107706*

4.6 Hydrothermal Process to Separate Blended Materials for Reuse in High Value Applications

Edwin Keh, The Hong Kong Research Institute of Textile & Apparel (HKRITA)

4.6.1 Introduction

Recycling and the reuse of old apparel and fabrics is a much-talked-about challenge. While we have the options to use our clothes longer, buy more second hand apparel, or give more of our old apparel to others, we are still eventually left with some unusable, damaged, and soiled garments. These need to be reprocessed or they will end up being landfilled or incinerated.

Today, these materials are recycled at very low rates. Low single digits are the most often quoted percentages. Of the materials reprocessed for reuse, most are downcycled to be used in lesser value applications. Possible uses include insulation, padding and rags.

The other problem we face with any type of reprocessing is the logistics challenge. Used materials are now mostly found in consuming economies and not in manufacturing economies. The 300-plus tons of daily discarded clothes from Hong Kong, for example, are generated in a city that no longer has yarn mills, fabric processing plants or apparel factories. The same is true across most of Europe and the Western economies. For the last 50 years, we have built a highly efficient, linear apparel-supply chain that makes apparel in the East for consumption in the West.

Most of our clothes today are made from blended materials. To reduce cost, improve comfort, and enhance performance, most are made from a blend of synthetic materials, cellulosic materials, and protein materials.

Polyester is today's most commonly employed material; it makes up more than half of all the materials used to make our clothes [1]. The second most common is cotton. Together with polyester, these two materials account for about 80% of all the materials in our clothes today. Polyester cotton blends are not only the most common blends in garments today, but also the fastest growing combination.

For over a decade now, excitement and hope have surrounded various chemical recycling processes as the way to dramatically increase the number of materials we can process and recycle. There is also optimism that a business case can eventually be made for a cost-effective chemical recycling industry.

Chemical recycling has the advantage of doing the minimum amount of damage to materials while maintaining the performance characteristics of the yarn and so producing a high-value output. There are numerous efforts around the world aimed at coming up with attractive solutions, but these have met with varying degrees of success so far.

A lot of the current chemical recycling processes involve the use of different processes to extract useful materials from textiles, yarns, and fibers. As the name implies, the use of chemicals in the process is a common theme. Operating conditions, however, may vary greatly. Most processes utilize chemicals, heat, pressure and some form of agitation as the conditions for reactions to take place.

Before materials can be chemically treated, there are the two issues of blend separation and dye removal to deal with. These two challenges currently are some of the biggest obstacles to scaling.

4.6.2 The Green Machine

In 2016, HKRITA and the H&M Foundation (HMF) worked together to find solutions for the separation and recycling of polyester/cotton blended materials. The first series of experiments with the H&M Foundation are commonly referred to as the "Hydrothermal Recycling System"[2]. This set of projects set about looking for practical ways to separate mixed synthetic and cellulosic fabrics. Separation usually involves degrading one material to harvest the other.

In most cases, this means polyester is chemically dissolved so that the cotton material may be recovered. Usually, the polyester then either is no longer useful or it is recovered in some depolymerized state. To reuse the polyester, the polymerization process has to be repeated. Often, there are additional processing steps for ensuring the quality, strength and performance of the recovered material. All this adds up to the use of more energy, chemicals, and investment in processing machinery.

While cotton is usually the more valuable of the two main materials, to an extent depending on the chemicals and processing conditions, it too may be recovered in a weakened state, in which event its value is lowered. Cotton material has to be disentangled, opened and carded in order for it to be reusable. Other chemical processes to treat the cotton are recovery of the cellulose material to make lyocell and other cellulose-based filament yarns and recovery of the cellulose as a pulp to be processed into new synthetic yarns.

Before materials can be chemically treated, dyes and other surface coatings have to be removed, as otherwise these may interfere with the chemical process. If the feedstock is from industrial sources, such as cutting waste, there should be good information available about the material composition and the dyes used. If the feedstock is from post-consumer sources, then there are significant unknowns.

---

[2] The hydrothermal experiments are known as ITP/103/15TP and ITP/025/17TP. For both these projects, the ITF was the majority contributor to the research funding.

In blended materials, there could be different dyes to remove. Reactive, disperse, and other types of dyes would need different types of processes. The good news is that there are already many solutions for this; however, each would create new waste streams to deal with, additional costs, and each would incrementally weaken or impair the quality of the output.

HKRITA's team looked at the various chemical solutions for separating the blended materials, but found that there are issues with the resulting chemical waste. Other possible separation methods explored also led to the deterioration of either the cotton or the polyester fibers and thus to lower economic yield of the outputs. The conventional recycling methods explored usually involve long chemical processes for depolymerizing and repolymerize the polyester, thus additing to the complexity and costs.

After several months of work, the HKRITA research team partnered up with a Japanese University to explore the possibilities of simplifying the chemistry involved in the separation process and began to work on using heat and pressure as the main mechanism for separating blended materials. It was found that, under the right operating conditions and with careful control, cotton fiber breaks down into short fibers and disentangles from the blend. The material can then be filtered out and dried, whereupon it forms a powder-like material. The key mechanism for effectively separating the blended material also turned out to be heat and pressure with minimal use of chemistry. This makes the process more cost effective and simpler and generates effectively no liquid waste.

4.6.3 How the Green Machine Works

The current hydrothermal process calls for the feedstock material to be prepared by cutting the fabric up into appropriately sized pieces. The material is then fed into a reactor vessel where heat and pressure are added. During the reaction process, some agitation is used to ensure a uniform level of heat, and pressure is maintained. After 1.5 to 3 hours, the reaction is stopped, the liquor is drained from the vessel and the feedstock is collected, filtered, and dried (Figure 4.29).

During the reaction, a large percentage of dyestuff and other surface coating is removed from the feedstock. Nevertheless, enough remains, necessitating a secondary process to remove the dye if the subsequent application calls for a light color; this is done either by running the material through activated charcoal filters or through the use of chemicals. These processes and the removed dyestuff and other coating materials are the key solid waste from this process. Currently, these are dried and disposed of, but their volume tends to be very small.

Polyester is recovered in fiber form. It can be used 'as is' as a staple. It can be reinforced by adding a percentage of virgin material if necessary. This material can be twisted together and used in knitted applications.

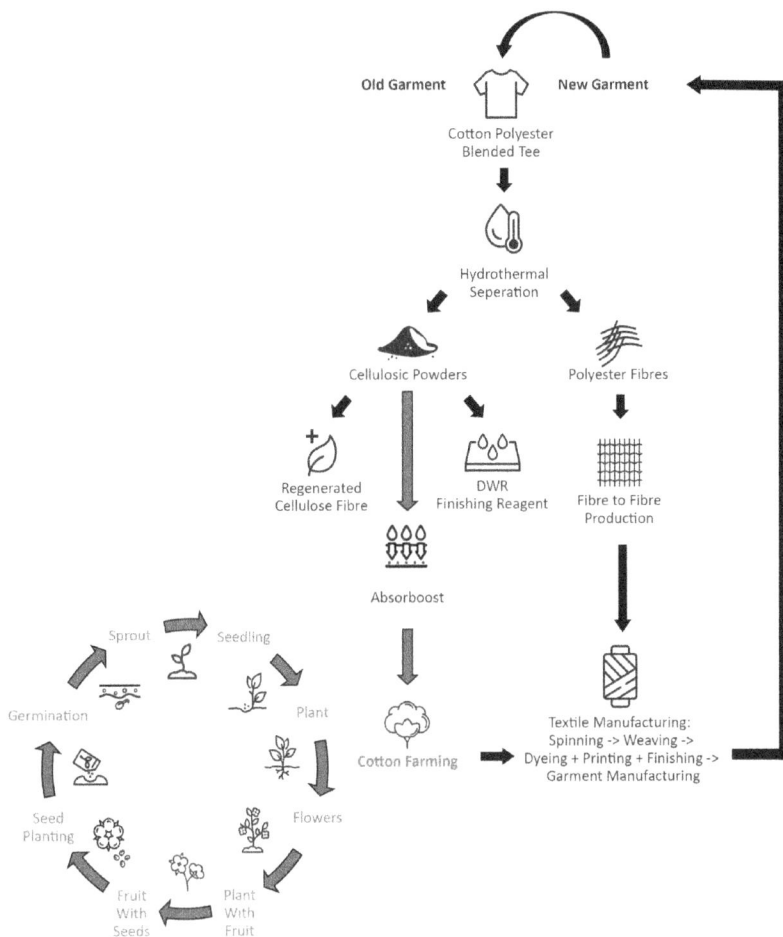

Figure 4.29 Current outputs from the Green Machine system

The polyester recovered exhibits no deterioration in quality and performance. This fact is attributed to the benign nature of the hydrothermal process. As no harsh chemicals, extreme heat or pressure are used, the polyester possesses the same properties as before the treatment. The possible uses for the separated polyester depend largely on what state it was in before the hydrothermal treatment. If the feedstock comprises post-industrial cutting waste, or stock in good condition, a minimal amount of treatment and blending are necessary for reuse. If the feedstock is generally worn, from post-consumer sources, or otherwise of poor quality, it is generally necessary to blend a percentage of virgin fibers to strengthen the material and improve the hand feel. Alternatively, the polyester output can be depolymerized and repolymerized to make new polyester filament. Naturally, here the challenge is the use of energy and chemistry. Another consideration when taking this approach is to ensure that filtration is done to a very high purity, with no contaminants.

Short cellulose fibers recovered from cotton are not long enough to be used 'as is' to make new fibers, so work is ongoing to find high-value applications for this. Our initial attempt involves using the cellulose in a new high-performance viscose yarn. The industrial scale process of this work is being conducted in partnership with Diawabo Rayon [2]. The goal here is to produce a new yarn from this recycled material that has performance characteristics. New patents have been filed by Diawabo in connection with this work.

Another work in progress is the use of the cellulose as a raw material for a new PRC-free DWR surface coating process [3]. Proof of concept and preliminary work on this process have been completed and are currently being refined.[3)]

An exciting application for the recycled cellulose obtained from the Green Machine is a new material that is currently being tested at scale for agricultural use. This new material is known as "Absorboost" or "Cellulosic Superabsorbant Polymer" or "C-SAP" [4]. Researchers at HKRITA crosslinked the recycled cellulose material to make a new polymer that possesses very good moisture-management properties. The Absorboost material can absorb moisture from the ambient environment and then release it upon saturation and repeat this process over a period of time. In ongoing cotton growing experiments in India, Absorboost is sprinkled around the roots of cotton plants during the growing season. With this application, cotton grows without the need for irrigation. Absorboost cotton plants are also at least 15% more productive in yield as a result. Experiments are ongoing to enhance the Absorboost with the addition of micronutrients to improve the health of the plant and improve the condition of the soil. At the end of the growing season, Absorboost material seems to naturally biodegrade, without leaving traces in the soil.

The experiments being conducted on Absorboost material are very exciting for HKRITA as there are lots of opportunities to improve cotton yield, to reduce water usage, and to potentially use it in other agricultural applications.

4.6.4   Green Machine Engineering & Scale-Up

There is ongoing work being done with the Green Machine. In 2017, with the hydro-thermal process breakthrough in HKRITA's labs, the HKRITA team set about looking at the engineering challenges of scaling a separation system to industrial scale. The first pre-industrial scale system was built in Hong Kong in 2018 (Figure 4.30).

---

[3] The PRC Free DWR surface coating work won an invention award, a special Gold Medal in 2021 in Geneva's Invention Exhibitions (*https://www.hkrita.com/tc/our-innovation-tech/achievement/awards/2021genevainventions/pfc-free-functional-Surface-finish*) as well as other scientific awards.

Figure 4.30 First pre-industrial scale system installed in Hong Kong in 2018

This small system has capacity to process about 150 kg of materials a day. It helped the researchers to work out the design of the processing tanks and the best way to use the least amount of energy for the reaction and to improve the recovery and reuse of the liquor. A tandem tank system was developed so that the reaction can take place in one tank while recovery and loading can take place in the other. Heated processing liquor is moved between the tanks to reduce heating energy. This process also optimized the productivity of the system and allowed the best operating conditions to be determined. It was found that, rather than build custom tanks for the hydrothermal system, conventional dye vats can be repurposed and retrofitted to become hydrothermal tanks. This translates to significant savings on the capital expenses of the system.

In 2020, all the engineering work resulted in the first industrial-scale system to be commissioned and built at a manufacturer in Bandung Indonesia (Figure 4.31).

Figure 4.31 Industrial-scale system installed in the Kahatex factory in Bandung, Indonesia

In the same year, the first capsule collection of clothes made with Green Machine fabrics was shipped. The used materials were made from the small Hong Kong pre-industrial scale system. The industrial scale system was completed in 2022. Due to travel restrictions in the middle of the global Covid pandemic, this took longer than expected. This new industrial-scale system has capacity to process up to 3 tons of material a day. The first fabrics and garments made from this system were shipped in late 2022.

In 2021, an agreement was signed with the denim supplier Isko for a Turkish-based system [5]. More systems are in various design and development phases around the world.

4.6.5 The Work & Opportunities Ahead

As interest grows more and more in both post-industrial and post-consumer fabric and garment recycling, there have been calls to develop Green Machine systems that are orders of magnitude bigger than the current system capacity. These are usually for systems with a capacity for 50 to 100 tons a day. Current constraints are the size of available dye vats that are repurposed for the hydrothermal process. The solution seems to be either to look at constructing larger-capacity purpose-built tanks or to develop multi-tank tandem systems that use conventional dye vats. Either path will lead to new engineering challenges.

Separation of protein materials is still an unsolved issue. Protein-based materials, such as wools and silks, are expensive and widely used. New processes are being worked on the separation and reuse of these.

System integration is also an opportunity. There are currently requests to use the Green Machine as part of integrated recycling solutions, especially in post-consumer applications. The challenge is to figure out how to position the feedstock and process the materials so that the highest value outputs are obtained . The Green Machine complements these efforts.

4.6.6 Conclusion

The research team at HKRITA feels that the matter of sustainability is not only time sensitive, it is also an existential threat, and critical. While the recycling of apparel may not be the silver bullet that will save our planet, the apparel, fashion, and textile industries are all significant contributors to pollution, produce significant greenhouse gases in production and transportation, and are significant consumers of resources, energy, and water. The sooner various recycling technologies are scaled up to industrial scale, the better.

The best way to scale solutions is to work as multi-disciplinary teams willing to take on some risks and simultaneously work on multiple scientific, engineering and business problems. This approach can accelerate progress and rapidly introduce solutions for industrial uses. However, there are significant financial and reputational risks in concurrently trying to solve system-engineering problems while still developing some of the necessary "science-at-scale" solutions. These include operational ergonomics considerations, system integration issues, materials logistics challenges, and new business model designs. All these issues, along with others yet to be considered, will have to be resolved in an accelerated time frame where possible. There are new complexities to manage and consider. However, this disruptive and less linear research model may support more rapid practical solutions to the huge sustainability challenges of the moment. The work on the Green Machine is a good example of how this may work.

References for Section 4.6

[1]   https://textileexchange.org/polyester/

[2]   https://daiwaborayon.co.jp/english/

[3]   https://www.hkrita.com/en/our-innovation-tech/projects/pfc-free-functional-surface-finish

[4]   https://www.hkrita.com/en/our-innovation-tech/projects/absorboost-csap-pilot-scale-system

[5]   https://www.just-style.com/news/isko-invests-in-hkrita-green-machine/?cf-view

4.7 Dissolution-Based Recycling Technologies for Textiles

Sea-Hyun Lee, Stefan Schonauer, Sascha Schriever, Institut für Textiltechnik of RWTH Aachen University, Germany

4.7.1 Cotton & Polyester – The World's Favorite Blend

To comprehend the prevailing fiber types in the market, it is essential to conduct an in-depth analysis of textile mill consumption. This analysis, along with rough estimates, provides indicators of the available resources. Global fiber demand for 2022 was estimated to be approximately 116 million metric tons (mt) [2], of which roughly 69% originated from crude oil as the base feedstock. Cotton accounted for 23% of world fiber demand, while only 6% was attributed to man-made cellulosic fibers (MMCF). A discernible trend emerges when synthetic fibers are dissected into their components. Polyester (PET) staple and filament fibers accounted for nearly 86% and 71% of synthetic fibers, respectively. Together, the cotton and polyester fractions, totaling the equivalent of 79.28% of global fiber consumption, constitute the most prevalent fiber blend used in today's apparel, home, and technical textile industry. Particularly in the apparel sector, the use of polyester and cotton is ubiquitous, due to their synergistic properties. Combined attributes such as material strength, durability, wrinkle resistance, color retention, versatility, and affordability cater to the properties required. With the advance of fast-fashion, polyester usage has surged substantially. While certain producers are transitioning to mono materials, which are ideal from a recycling perspective, most companies opt for blends with varying ratios. Common cotton/polyester ratios on the market are 80/20, 65/35, 50/50, 35/65, and 40/60. Ratios such as 50/50, 65/35, and 35/65 are typical for everyday clothing whereas, for sleepwear, anti-allergic apparel, and workwear, the proportion of cotton markedly increases. Sports clothing and uniforms, which demand higher fiber performance, often incorporate a higher polyester content, with a ratio of 40/60 commonly employed. In addition to these common blends, other options include polyester, cotton, viscose (50/25/25), and polyester and elastane blends (85/15) [3–4].

An intriguing observation is linked to the regional utilization of cotton and polyester blends. More-developed countries in the European Union, Asia, and North America tend to use a higher concentration of cotton than emerging economies, which rely heavily on PET [1].

While most technologies currently focus on pre-consumer waste, the ability to process post-consumer waste will become the next crucial step toward closing the loop and achieving circularity. However, current recycling rates for cotton only account for 1% of the 25 million tons of virgin material [2]. Since technologies, such as bottle-to-fiber, are soon to be phased out as food-grade plastic recycling to fiber is now considered a down-grade, new sources are needed for the textile recycling industry [5]. With, for example, Germany mandating the separation of old textiles by law in 2025, the push to textile fiber-to-fiber recycling with increasing feedstock mass streams would seem to be growing more viable [6].

Pre-/post-consumer textile waste is defined by K.H. Tang [7] as follows:

- Pre-consumer textile waste includes production waste generated during the processing of fibers, yarns, textiles, technical textiles, and non-wovens, including off-cuts, selvages, and rejected materials. Pre-consumer waste is considered to be clean and of high quality.

- Post-consumer textile waste consists of textiles that consumers have disposed of. The quality of recovered material subject to recycling varies with the recovery system.

- Industrial textile waste is predominantly generated from commercial applications, including carpets and curtains. Industrial waste streams are considered to be contaminated and of low quality.

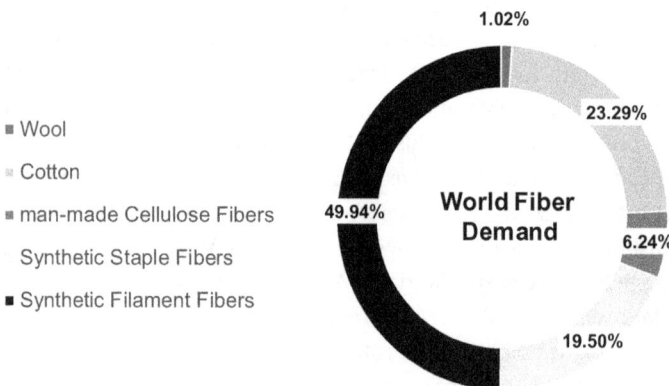

Figure 4.32 Overview of the polyester and cotton supply chain for textiles

4.7.2 Dissolution Technologies

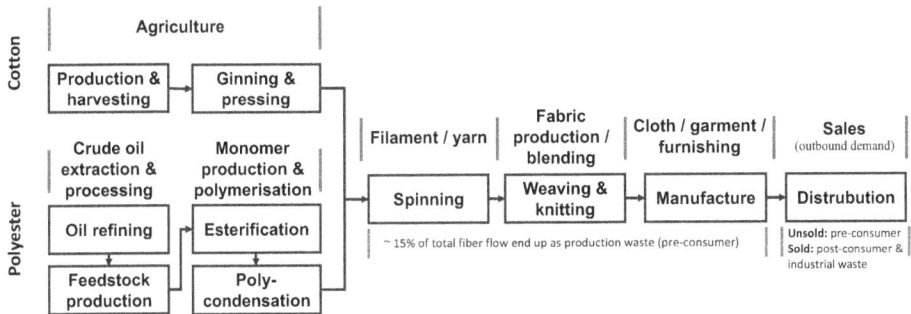

Figure 4.33 Chemical & dissolution recycling routes

Recycling technologies for fibers vary significantly with the desired product and type. They can be categorized into mechanical, dissolution, depolymerization, and conversion processes. Figure 4.33 shows a conventional route for textiles with added recycling processes, with the processes split into the target output products, which are marked in orange. While mechanical recycling is commercially available, chemical and conversion technologies have yet to demonstrate viability on a large scale. Moreover, the efficacy of certain technologies depends on the material being utilized. Presently, depolymerization of textiles is primarily feasible for polycondensates, such as PET and polyamides. Positioned between chemical and mechanical recycling, a novel technology known as dissolution has emerged that combines favorable properties of both methods. Dissolution technologies can effectively separate fiber blends in a manner akin to chemical recycling, but without degrading the polymer into its monomer state, and thereby simplifying the application of the target product within existing platforms. [12]. Dissolution offers greater flexibility, as it can dissolve PET, polyacrylonitrile, polypropylene, polyethylene, polystyrene, and polyvinyl chloride. Ideally, the target polymer retains its properties, yielding a material of near-virgin quality, although some degree of degradation may occur. This poses challenges for such polymers, as a certain amount of virgin polymer is still necessary for maintaining essential properties [13]. However, for cotton, dissolution presents a unique advantage since its primary component is cellulose, the foundation of manmade cellulosic fibers (MMCF). With an average degree of polymerization between 300–700 dp for MMCF, cotton requires partial degradation into an ideal polymerchain length, the so-called aging for stable wet spinning processes, indicating degradation of cotton during the recycling process generally favorable from a processing standpoint [14].

The primary question then becomes which polymer should undergo depolymerization or dissolution, as both cotton and polyester are technically viable options for these processes.

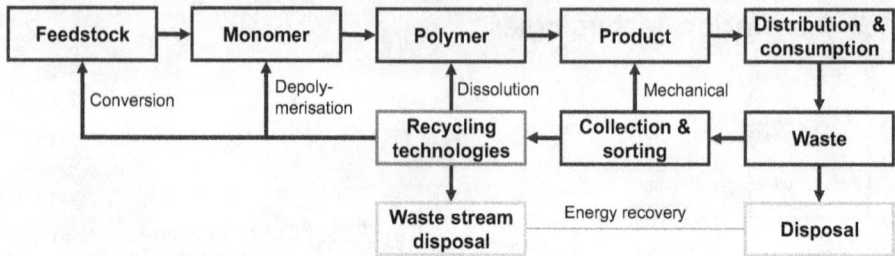

Figure 4.34 General recycling overview for dissolution technologies of cellulose and PET-based textiles

Table 4.1 Example (Co-)Solvents for PET Dissolution [15]

Solvents	Hazard	Health	Environment
N-Methyl-2-Pyrrolidone (NMP)	1	9	7
Dimethylacetamide (DMAc)	1	9	5
DMSO	1	1	5

In cases where fibers meet the minimal requirements for the production of new man-made cellulosic fibers, complete breakdown of the material can be avoided. In such instances, dissolution and adjustments of the polymer chain suffice. There is a wider array of options for cellulose dissolution. In cellulose fiber production, possibilities include derivatization or direct dissolution. Derivatization is utilized in the viscose process to form xanthogenate for dissolution. Direct dissolution is prominent in the lyocell process, where an organic solvent like N-methylmorpholine-N-oxide (NMMO) can directly dissolve cellulose in its native state, bypassing the need for the derivatization required in other processes. Besides viscose and lyocell, other fiber-spinning technologies, such as carbacell and cupro, and salt hydrate melts, such as zinc chloride, can also act as solvents for cellulose. While additional methods exist, their effectiveness on a commercial scale in combination with fiber spinning remains to be demonstrated.

Table 4.2 Common Solvents for Cellulose [16]

Common Solvents	Textile Connection
Sulfuric acid	Commercial (needed for Rayon)
Sodium hydroxide/urea	Commercial carbacell
Sodium hydroxide /zinc oxide	Variant of carbacell
Tetraammine copper [II] hydroxide	Commercial cupro
Zinc chloride (salt hydrate melt)	Lab scale

Common Solvents	Textile Connection
NMMO	Commercial (lyocell)
N,N-dimethylacetamide (DMAc)/ lithium chloride (LiCl)	Lab scale
Tetrabutylammonium fluoride (TBAF) / acetate / DMSO	Lab scale

Other options under consideration for commercialization are a relatively new class of solvents known as ionic liquids (IL). ILs are a type of salts with melting points below 100 °C and are classified as green materials. They are highly effective; however, they are costly, and high levels of recycling are need in order for the process to be economically viable. These processes typically require higher specifications for the quality of the input material. Therefore, ensuring the recyclability and degradation, along with managing contaminants, is crucial for commercial application in the recycling industry. Particularly with post-consumer goods, efficient solvent recycling systems have been studied and are listed in Table 4.3.

Table 4.3 Ionic Liquids for Cellulose Dissolution [17–18]

Ionic Liquid	Cation	Anion	Co-Solvent
Alkyl-imidazolium cations	[Cn-mim]	[Cl]	DMAc
Pyridinium-based	[Cn-Pyr]	[Br] [SCN]	DMSO DMF
Ammonium-based	[TBA] [ETOHA]	[HCOO] [CH3COO]	
Phosphonium-based	[Pn]	[CH3CH2COO], [C6H5COO]	
Morpholinium-based	[Cn-Mmorf]	[HSCH2COO] [(MeO9)RPO2]	
Tetramethylguanidium-based	TMGH	[(RO)2PO2]	
1,5-Diazabiclyclo[4.3.0] non-5-enium-based	DBNH		
1,7-Diazabicyclo [5.4.0]undec-6 ene-based	DBUH		

From the field of ionic liquids, a new class of solvents has garnered significant interest in the scientific community, namely deep eutectic solvents. These are generally easy to handle, low-cost, and safe to use. However, as of now, no commercial applications have been found.

4.7.2.1 Example: Re:NewCell

One of the current innovators in the field of textile recycling is Re:NewCell, a Swedish company that commenced commercial production in 2022 with a capacity of 60,000 tons per annum in Sundsvall, Sweden. Their processes include shredding, de-buttoning, dezipping, and de-coloring, after which the cellulosic fibers are dissolved into a slurry. A closer examination of Re:NewCell reveals a process involving an alkali solution combined with textiles, pressurized with oxygen and/or another gas, occurring below room temperature [19]. It is assumed that this process indicates the hydrolysis of cellulose into spinnable length. Precipitation follows through addition of at least one organic solvent and final separation. The cellulose is then formed into sheets and dried, creating a dissolving grade pulp from 100% recycled textiles for further processing into viscose fibers. Fibers from Re:NewCell are comparable to virgin viscose staple fibers and cotton in terms of titer and tenacity. Consequently, they are used in the apparel sector by various well-known brands, such as Levi's, Zara, H&M, GANNI, and Pangaia. Re:NewCell recycled fiber properties are shown in Table 4.4 [20].

Table 4.4 Comparison of Virgin Fibers Versus Recycled Fibers

Fiber Type	Recycled Content	Titer [dTex]	Tenacity [cN/ dTex]
Re:NewCell viscose staple fiber	30–50%	1.67–1.33	> 2
Example: viscose staple fiber	0%	1.45	2.5
Example: cotton	0%	1.6	2.79

References for Section 4.7
[1] Murphy, L., Global Fibers Strategic Planning Outlook 2023, Wood Mackenzie, 2023, Report

[2] Textile Exchange, Market Market Report 2023, Textile Exchange, 2023, Report

[3] Nguyen, B., Cotton Polyester Blend Explained: Pros,Cons, Applications, Merchize, 2024, Article (accessed: 25.03.2024)

[4] Textile School, Blended Fabrics, Textile School, 2018, Article (accessed: 25.03.2024)

[5] European Parliament, REPORT on an EU Strategy for Sustainable and Circular Textiles, European Parliament, 2023 (A9-0176/2023) (accessed: 25.03.2024)

[6] Umweltbundesamt, Rechtsgrundlagen Europäisches Recht, Abfallrecht, 2022, Website (accessed: 18:39; 25.03.2024)

[7] Tang, K. H., State of the Art in Textile Waste Management: A Review. Textiles, 2023, 3, 454–467

[8] Gries, T., Veit, D., Textile Technology. An Introduction, 2015, Hanser, Munich

[9] Palacios-Mateo, C., van der Meer, Y., Seide, G., Analysis of the polyester clothing value chain to identify key intervention points for sustainability, Environmental Sciences Europe, 2021, 33, 2

[10] Hedrich, S., Janmark, J., Strand, M., Langguth, N., Magnus, K.-H., Scaling textile recycling in Europe-turning waste into value, McKinsey, 2022, Article

[11] Lange, J.-P., Managing Plastic Waste – Sorting, Recycling, Disposal, Product Redesign, ACS Sustainable Chem. Eng., 2021, 9, 47, 15722-15738

[12] Vollmer, I., Jenks, M.J.F., Roelands, M.C.P., White, R.J., van Harmelen, T., de Wild, R., van der Laan, G.P., Meirer, F., Keurentjes. J.T. FB.M. Weckhuysen, B.M., Beyond Mechanical Recycling: Giving New Life to Plastic Waste, Angewandte Chemie, 2020, 59, 26, 15402-15423

[13] Hann, S., Connock, T., Chemical Recycling: State of Play, Eunomia Research & Consulting, 2020, Report

[14] Blanco, A., Monte, M., Campano, C., Balea, A., Merayo N., Negro, C., in: Handbook of Nanomaterials for Industrial Applications, 74–126

[15] Byrne, F. P., Jin, S., Paggiola, G., Petchey, T., Clark, J., Farmer, J., Hunt, A., McElroy, C., Sherwood, J., Tools and techniques for solvent selection: green solvent selection guides, Sustainable Chemical Processes, 2016, 4,7

[16] Sirviö, J. H., Heiskanen, J. P., Room-temperature dissolution and chemical modification of cellulose in aqueous tetraethylammonium hydroxide-carbamide solutions, Cellulose, 2020, 27, 1933–1950

[17] Holding, A., Parviainen, A., Kilpeläinen, I., Soto, A., King, A., Rodriguez, H., Efficiency of hydrophobic phosphonium ionic liquids and DMSO as recycleable cellulose dissolution and regeneration media, RCS Advances, 2017, 7, 17451-17461

[18] Zhang, J., Wu, J., Yu, J., Zhang, X., He, J., Zhang, J., Application of ionic liquids for dissolving cellulose and fabricating cellulose-based materials: state of the art and future trends, Materials Chemistry Frontiers, 2017, 1, 1273–1290

[19] Lindström, M., Henriksson, G., Re Newcell AB, Regeneration of Cellulose, European Patent Office, EP2817448B1, 16.11.2016

[20] Re:NewCell, Annual and Sustainability report 2022, 2022, Report

5

Strategies for Transforming to a Circular Economy

5.1 Business Models for a Circular Textile Economy

Roxana Ley, Lukas Balon, Institut für Textiltechnik of RWTH Aachen University, Germany

5.1.1 Market for Fiber Production and Recycling

It is becoming increasingly clear that the current linear economic model (take-pro-duce-dispose) is no longer sustainable. Fortunately, the principles of the circular economy, which are based on the continuous recovery and regeneration of value in the life cycle of products, are increasingly being recognized in Europe. The circular economy model follows the 3R principles: Reduce, Reuse, Recycle, which are applied throughout the life cycle of products. The circular economy aims to optimize the use of existing resources and create renewable material and product flows.

5.1.1.1 Fiber Production

At present, around 113 million tons of natural and man-made fibers are produced each year. Added to this are glass and metal fibers, so that the total market amounts to over 130 million tons. In 2020, approximately 32 m tons of natural fibers were produced, of which cotton is by far the most important. This contrasts with 81 m tons of man-made fibers on a synthetic and cellulosic basis. Polyester (PET) accounts by far for the largest share, mainly as filament [1].

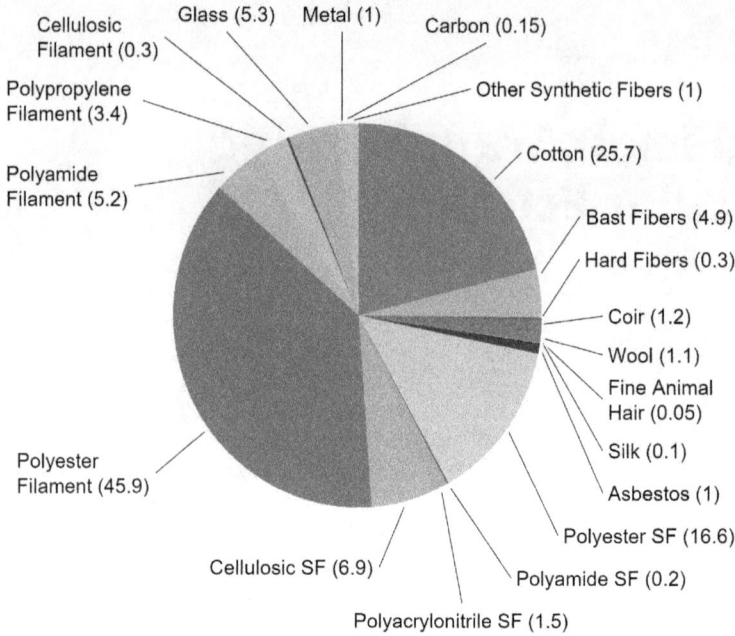

Figure 5.1 Share of important fibers in the global market [million t] [1]

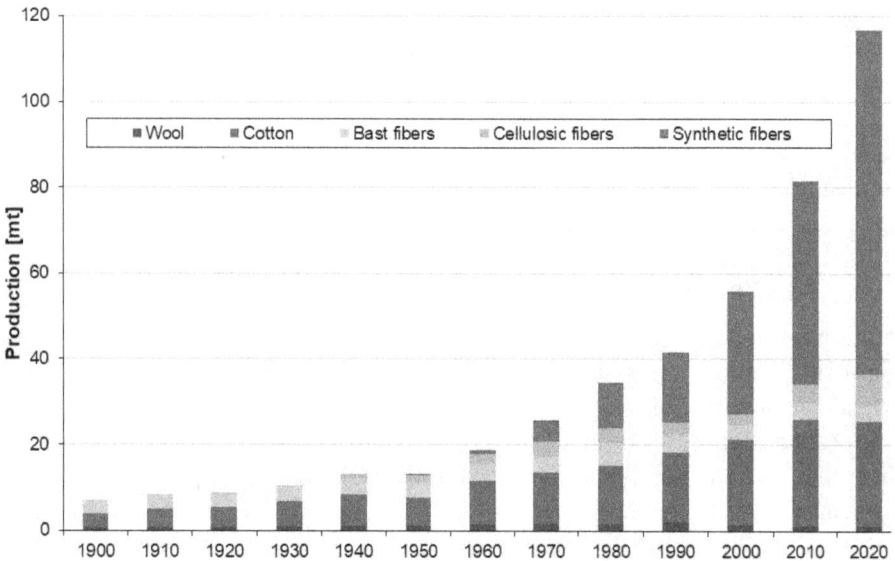

Figure 5.2 Share of important fibers in world production [1]

The production of man-made fibers on a larger scale did not begin until around 1900. Until 1960, man-made cellulosic fibers dominated, followed by polyamides from 1950 onwards, before the production of polyester, especially PET, increased massively since

1970. In 1992, PET overtook cotton as the most important fiber in terms of volume and, in 1997, the volume of man-made fibers produced exceeded the production of natural fibers. After many years of stagnation, since 2010 the production of cellulosic-based fibers, mainly viscose and lyocell, has been rising steeply, by 64% from 2010 to 2020. The relative increase in the production of synthetic fibers during this period is lower, at 57%. It is expected that the importance of cellulosic fibers will continue to rise because they are obtained from renewable raw materials and are therefore considered more sustainable than petroleum-based synthetic fibers [1].

5.1.1.2 Market for Fiber Recycling

The global textile recycling market size was valued at 4,632.4 million US dollars in 2022 and is expected to grow at a compound annual growth rate (CAGR) of 3.2% from 2023 to 2030. The rising environmental concern about textile waste production and growing social awareness about textile recycling is expected to propel the market growth. According to the Environmental Protection Agency (EPA), 5% of landfill space is occupied by textile waste [2].

Europe region led the textile recycling industry and accounted for 29.6% of the global revenue share in 2022. The market in the region is also expected to grow significantly in the coming years owing to surging support from governments of the countries and an increasing number of initiatives related to recycling. Leading European countries for textile recycling are Germany, Italy, France, Belgium, the Czech Republic and Poland [2].

5.1.2 Circular Economy

Circular economy is a production and consumption model that aims to maximize the useful life of materials and products through strategies such as sharing, leasing, reusing, repairing, refurbishing and recycling. The goal is to extend the life cycle of products and reduce waste to a minimum. In contrast to the traditional linear economic model, which is based on the principle of the "throwaway economy" and is characterized by high quantities of cheap and easily accessible materials and energy, circular economy strives to keep resources in the economy for as long as possible. This is done by dismantling products at the end of their life and reusing their components to enable continuous value creation [3].

In terms of environmental protection, circular economy slows down the use of natural resources by reusing and recycling products. This reduces the destruction of landscapes and habitats and contributes to the preservation of biodiversity. It also reduces greenhouse gas emissions: industrial processes and product use account for 9.10% of EU emissions, while waste management accounts for 3.32%. Efficient and sustainable product designs could significantly reduce energy and resource consumption, as over 80% of environmental impacts are determined at the design stage. Switching to durable, reusable products would also reduce the amount of waste [3].

Given the limited availability of key raw materials and growing demand, due to population growth, reducing dependence on raw materials is becoming increasingly important. The EU imports around half of the raw materials it consumes and recorded a trade deficit of 35.5 billion euros in 2021. By recycling raw materials, supply risks such as price fluctuations and import dependency can be mitigated, especially for critical raw materials that are necessary for climate-relevant technologies [3].

The transition to a circular economy could increase competitiveness, promote innovation, boost economic growth and create 700,000 jobs in the EU by 2030. Redesigning materials and products for circular use will stimulate innovation in different sectors of the economy. Consumers will benefit from more durable and innovative products that improve quality of life and save costs in the long term [3, 4].

5.1.3 Greenwashing

According to the Cambridge Dictionary, greenwashing is defined as "behavior or activities that make people believe that a company is doing more to protect the environment than it really is" [5].

A prominent example of greenwashing is self-assessed sustainability labels, which are often used by brands for marketing purposes. In a review conducted by Greenpeace, a total of 29 marketing labels from different brands were evaluated. During this review, several patterns of concern emerged, including:

- Confusing consumers with tags which are featured as if they were certified labels and are sometimes named after company sustainability programs.

- A lack of third-party-verified or in-house evaluation of compliance with the best available standards on the environment, social and human rights.

- A lack of supply chain traceability beneath the label.

- The continued production of fiber blends, such as polycotton, which are presented as greener, due to their recycled content, despite the fact that mixed fibers are a one-off unrecyclable solution that do not close the loop.

- Not providing consumers and third parties with a breakdown of figures per material to substantiate the company's green claims or its overall direction and long-term strategy.

- Some labels highlight a single aspect of improvement in production, such as the reduction of water use or the reuse/recycling of pre-consumer waste.

- However, communication by brands about sustainability is not only limited to the use of marketing labels on products. Brands also often use subliminal ways of promotion, such as in-store advertising [6].

5.1.4 Circular Business Models

Business models play a crucial role in applying the concept of circular economy at the organizational level. Despite the growing body of literature exploring circular business models and innovations in this area, there remains significant ambiguity in their theoretical frameworks. This is reflected by a range of definitions for the term "circular business model." Based on a literature review, Geissdoerfer et al. suggest the following definition:

"Circular business models can be defined as business models that are cycling, extending, intensifying, and/or dematerializing material and energy loops to reduce the resource inputs into and the waste and emission leakage out of an organizational system. This comprises recycling measures (cycling), use phase extensions (extending), a more intense use phase (intensifying), and the substitution of products by service and software solutions (dematerializing)" [7, 8].

On the basis of this definition, four generic strategies for circular business models can be distinguished. These strategies, alongside implementation examples, are described in Table 5.1.

Table 5.1 Strategies for Circular Business Models, Adapted from [6]

Strategy	Description	Implementation Examples
Cycling	Materials and energy are recycled within the system, through reuse, remanufacturing, refurbishing, and recycling	⬚ Reuse ⬚ Repair ⬚ Recycling ⬚ Refurbishing
Extending	The use phase of the product is extended, through long-lasting and timeless design, marketing that encourages long use phases, maintenance, and repair.	⬚ Long-lasting products ⬚ Upgradability ⬚ Timeless design ⬚ Maintenance, Product Support
Intensifying	The use phase of the product is intensified through solutions such as sharing economy	⬚ Sharing models ⬚ Renting models ⬚ Pooling models
Dematerializing	Product utility is provided without hardware through substitution with service and software solutions	⬚ Software instead of hardware ⬚ Service instead of product

The implementation of these strategies in the context of textiles will be discussed in the following section.

5.1.4.1 Recycling

During use, a product is subjected to wear and tear, which means that it is no longer fit for purpose after a certain period of time. Suitable care, such as regular washing of textiles, can extend the life of the product. However, the limit of usability is determined by the cost-benefit ratio and the cost of purchasing a new product. In Germany, around 2 million tons of new textiles are placed on the market every year. This quantity includes not only used textiles, but also new textiles that do not reach the market due to manufacturing defects. As the same quantity of textiles has to be disposed of, the question of suitable recycling methods arises. Textile recycling refers to the process in which used or defective textiles are recycled using various methods in order to conserve resources and reduce waste. There are three options available for this: material recycling, chemical recycling and thermal recycling.

Mechanical Recycling

During the mechanical processing of textiles, they are first shredded into small fragments. Particularly in the processing of synthetic base materials, such as PET bottles, the fragments are then melted, followed by pressing through a special device to enable the formation of a new yarn. If necessary, new fibers can be added to improve the quality of the recycled material if it cannot match the performance of virgin materials. This process approach is not only used in the recycling of PET, but is also particularly suitable for the reprocessing of cotton, as the characteristics of the cotton fibers are retained during the process. However, it should be noted that only around 20% of the material used can be effectively recovered and made usable for further processing. The quality of the recycled cotton material often tends to be lower than that of new cotton fabric, due to the shortened fiber lengths of the recycled material. For this reason, virgin cotton fibers, i.e. freshly harvested cotton fibers, are often added to the recycled material to improve its quality. It should also be noted that, despite its advantages, the cotton recycling process also presents some challenges [9].

Through thermal treatment, fibers can be obtained from PET bottles that can be used for the production of textiles. The steps of the process are:

- Removal of impurities
- Shredding of the bottles
- Melting with an extruder
- Extraction of the fibers

This process usually leads to the production of nonwovens instead of yarns, as the recycled PET raw material often no longer has the required properties for yarns after melting, such as adequate chain length. The economic viability of this process depends on whether the PET bottles are available at low cost and whether a high price can be achieved for the PET nonwovens produced from them [10].

Among other things, mechanical shredding and reprocessing weaken the structure of the cotton fibers and can lead to a reduction in their strength and longevity. This effect is exacerbated by the need to add additional virgin cotton fibers to compensate for the deficiencies in the recycled material. Furthermore, the separation of cotton fibers from other materials, such as polyester or spandex fibers, is a technical challenge that can affect the efficiency of the recycling process. Research and development is therefore aimed at developing improved processes and technologies to optimize the quality and sustainability of the recycled cotton material and increase the proportion of recovered material [11].

Table 5.2 Advantages and Disadvantages of Mechanical Recycling

Advantages	Disadvantages
Energy efficiency: due to the less complex process and lower temperatures	**Loss of quality:** fibers are shortened or damaged and do not have the same performance or durability as the original material
Maintenance of material properties: original properties of the materials can be largely retained	**Limited range of materials:** similar fiber-composition must be pre-set, mixed or complex materials may not be recycled effectively

In the context of mechanical recycling – particularly for plastics and textiles – new business models (NBMs) can play a pivotal role in implementing sustainable practices while simultaneously realizing economic benefits. A traditional business model within mechanical recycling involves the production of lower-quality goods from recycled materials. This model is based on the concept of cascading use, where materials that are no longer suitable for their original purpose are repurposed into new, lower-grade products. An example of this would be the production of insulation materials from recycled plastics or textile fibers. The advantages of this model lie in the cost-effective manufacturing process and the reduction of waste streams and the simultaneous tapping of new markets for lower-grade materials.

An innovative business model idea within the mechanical recycling sector could involve the integration of used-clothing collection containers with an in-built shredder. These containers would enable the direct shredding of old clothes on-site, offering several advantages. Firstly, the logistics chain would be significantly streamlined, as the shredded textile fibers would take up less volume and be cheaper to transport. Secondly, the on-site shredding could improve the material separation process, as the fibers could be sorted and processed at an earlier stage.

Chemical Recycling

Chemical recycling is primarily employed for synthetic fibers such as polyester. This process commences with the fragmentation of textiles, followed by their conversion into pellets or flakes. These fragmentations facilitate the return of synthetic materials to their basic state, subsequently reconstituting them into yarns or textile surfaces. The chemical breakdown of synthetic fibers necessitates significant energy expenditure, as high temperatures are required to rupture molecular bonds and isolate constituents. Another crucial aspect is the condition of the recycled materials, which substantially influences the efficiency of the recycling process. Particularly significant are factors such as the pretreatment of old textiles, e.g. through dyeing or chemical treatment. The removal of additives, such as dyes, bleaches, or softeners, is a time-consuming and cost-intensive process. However, the realm of chemical recycling is currently experiencing notable dynamism: companies, such as Rittec from Lüneburg, are introducing innovative technologies that provide an effective textile recycling solution for synthetic waste textiles. Rittec emphasizes that their technology enables a reduction of up to 40% in CO_2 emissions compared with conventional polyester production from crude oil. Additionally, the chemical company Carbios, in collaboration with leading textile companies, has established a recycling consortium that is collectively researching solutions and addressing the current challenges facing the sector [12].

In the chemical reprocessing of cotton, the specific material properties of the fabric are dissolved, yielding pulp. This pulp can be utilized in various applications, such as insulation material [13].

Table 5.3 Advantages and Disadvantages of Chemical Recycling

Advantages	Disadvantages
Maintenance of material properties: material properties of the source materials can be largely retained	**High consumption of energy and resources:** high amounts of energy and chemical additives to process the materials
Wide range of applications: chemically recycled materials can be used from the production of new textiles to use in other industries	**Complexity and costs:** specialized equipment, technologies and expertise are required

Integrating chemical recycling into the European textile value chain offers the opportunity of reindustrializing raw material production within Europe; a significant amount of that production has migrated to Asia in recent decades. By developing and implementing an efficient chemical recycling technology, this approach seeks to promote a circular economy, reduce reliance on imports and enhance sustainability within the European textile industry. Core components of this approach are:

- **Technological Innovation and Scaling**: Developing and optimizing a cost-effective chemical recycling technology specifically tailored to textile waste

- **Reindustrialization and Location Strategy**: The production process will be relocated to European countries with a strong industrial base and the necessary infrastructure. Proximity to sources of textile waste and downstream industries will be crucial for minimizing transportation costs and bolstering the local economy.

- **Economic Viability and Market Potential**: Repatriating raw material production to Europe to lower import costs and create jobs. The growing demand for sustainably produced textiles expands the market for recycled raw materials, ensuring steady revenues over the long term

Thermal Recycling

Thermal recycling, also known as energy recovery, is becoming increasingly important in the debate on sustainable waste management. In this process, textiles are thermally treated in special incineration plants. The process involves the controlled incineration of textile waste, which releases thermal energy. This energy can then be used in the form of heat or electricity to support other industrial processes. Compared with mechanical and chemical recycling, thermal recycling offers an additional way of harnessing the energy contained in waste materials. While mechanical recycling aims to physically reuse materials and chemical recycling changes the molecular structure of the polymers, thermal recycling aims to release the energy stored in the materials and make it usable. This typically takes place in incineration plants with energy recovery systems, which enable a reduction in overall energy consumption and a reduction in greenhouse gas emissions through the efficient use of the energy generated. The thermal recycling of textiles therefore not only helps to reduce the amount of waste generated, but also supports the circular economy by enabling waste materials to be put to good use. However, it is important to note that thermal recycling is considered less desirable in the waste hierarchy than the reuse or recycling of materials. Nevertheless, it plays a crucial role in a comprehensive waste management system, especially for materials that are unsuitable for other recycling methods.

Table 5.4 Advantages and Disadvantages of Thermal Recycling

Advantages	Disadvantages
Efficient material transformation: different materials can be burned together	**Environmental impact:** release of pollutants and greenhouse gases
Reduction of landfill waste: reducing the volume of waste and thus reducing the environmental impact	**Energy consumption:** the use of high temperatures increases the energy requirement, which increases the environmental impact

One possible application for thermal recycling of textiles is the provision of electricity and process heat for industrial processes, for example in the chemical industry. This approach enables the efficient utilization of textile waste, reduces landfill burden and contributes to a circular economy by maximizing the energy recovery from waste. Additionally, replacing traditional gas consumption in industrial processes with energy from textile combustion can slash the overall demand for natural gas.

5.1.4.1.1 Downcycling

Downcycling is a recycling process where the value of the recycled material decreases over time, being used in less valued processes with lower-quality material and with changes in inherent properties, when compared with its original use. Reasons for downcycling include a poor design of products which does not take recycling or disassembly into account and inadequate end-of-life management of products and materials, leading to contamination with other substances and materials and thus a material of low quality. This significantly limits the applications of those materials [14].

Generally, downcycling can be seen as a cascade system as, with every recycling step, the value of the product decreases. Eventually, products with such little value are produced, that even downcycling is not economically viable anymore. These products are then either burned to recover energy or stored on landsides.

The downcycling of textiles is often closely linked to mechanical recycling, since these processes often lead to a significant reduction in fiber length. Thus, typical examples of downcycling include insulation materials or cleaning cloths.

5.1.4.1.2 Upcycling

Upcycling was first introduced by McDonough and Braungart as "the practice of taking something that is disposable and transforming it into something of greater use and value" [14]. Upcycling is generally understood as a design-based circular fashion approach, where pre- or post-consumer textile waste material is repurposed to create new garments [15]. Thus, upcycling can be seen as the opposite of downcycling.

5.1.5 Reuse

Reuse plays a central role in textile recycling and makes a significant contribution to reducing textile waste. Instead of textiles being disposed after use, they are reprocessed through various processes and reintegrated into the utilization cycle. This primarily involves the resale of used garments. Reuse offers ecological benefits by reducing the need for new production and thus minimizing the consumption of resources and the environmental impact of textile production. In addition, reuse promotes social sustainability by creating jobs in the sorting, processing and marketing of used textiles. Innovative business models, such as clothes swaps and second-hand platforms, also help to promote reuse and raise consumer awareness of sustainable consumption habits [16].

In the realm of new business models (NBMs), reuse plays a pivotal role, particularly through the establishment and promotion of second-hand markets. Second-hand retailing represents the most effective form of reuse, as it significantly extends the lifespan of products, thereby realizing both ecological and economic benefits. However, for these models to be successful and sustainable, certain requirements must be met: high product quality, efficient sorting, and prioritization of the durability of goods.

Quality Assurance:

A successful second-hand business model is fundamentally built on the high quality of the items being resold. Only products in good condition can earn the trust of consumers, who are increasingly discerning, even when purchasing used goods. By maintaining rigorous quality standards, it is possible to ensure that products can undergo multiple reuse cycles, which in turn enhances the sustainability of the model. High-quality items not only attract better prices, but also strengthen the reputation of the second-hand market, making it more appealing to a broader audience.

Efficient Sorting and Categorization:

Efficient sorting processes are essential for maximizing the value of second-hand goods. Items must be categorized not only by type, but also by quality, brand, and potential resale value. Advanced sorting techniques, potentially supported by automated systems, can significantly enhance the efficiency and accuracy of this process. Effective sorting is crucial to ensuring that only marketable items are selected, while those that do not meet the required criteria can be redirected to recycling or upcycling streams. This approach reduces waste and optimizes the flow of goods within the business model.

Durability and Longevity:

Another key factor in the viability of second-hand markets is the durability and longevity of the products being resold. Items designed for long-term use are more likely to retain their value over time, making them particularly suitable for the second-hand market. Therefore, the success of reuse-based business models depends on both manufacturers and consumers prioritizing durability as a core design and purchasing criterion. This focus not only supports reuse, but also contributes to broader sustainability goals.

5.1.6 Repair and Maintenance

Repairing describes all efforts to "put something that is damaged, broken, or not working correctly, back into good condition or make it work again" [17]. As such, repairing is commonly seen as an integral part of a circular textile economy. Advantages and disadvantages of textile repair are summarized in Table 5.5:

Table 5.5 Advantages and Disadvantages of Textile Repair

Advantages	Disadvantages
Environmental benefits: repairing clothing reduces waste and conserves resources and energy needed for the production of new products	**Effort**: repairing clothing requires time and effort, which may not be appealing for some consumers
Value retention: unlike downcycling, high-value products can be used for a longer time, thus retaining their value	**Skill and equipment**: repairs often require special skills and equipment that not all consumers possess

Table 5.5 Advantages and Disadvantages of Textile Repair (*continued*)

Advantages	Disadvantages
Skill development: repairing clothing can be a valuable skill and promote self-reliance	**Availability**: access to professional and affordable repair services varies with location and financial situation
Job creation: on a larger scale, the development of a repair economy can create new jobs	**Expenses**: professional repair services can be more expensive than buying new, particularly for inexpensive pieces of clothing
Personalization: repairs allow for personalization of clothing	**Durability**: repairs may not always restore the original properties, thus negatively affecting the durability and functionality of the product
	Aesthetics: repairs can detract from the appearance of clothing

An increasing number of fashion brands are actively encouraging the repair of their products. These efforts may include repair services, repair instructions or DIY kits. Examples of repair programs are summarized in Table 5.6. It must be noted that this selection does not constitute a judgement on repair programs offered by other fashion brands.

Table 5.6 Examples of Repair Programs Offered by Different Fashion Brands

Fashion Brand	Repair Efforts
Patagonia	Patagonia's Worn Wear program includes repair guides, repair services and the option to buy used Patagonia gear.
Nudie Jeans	Nudie Jeans offers a free repair service for their jeans, which includes repair shops and repair partners around the world.
Arc'teryx	Arc'teryx has a comprehensive repair program that offers repairs for their gear. If a product can no longer be repaired to quality standards, it will be donated to charities that encourage outdoor activities.
The North Face	The North Face Renewed is a resale marketplace for buying returned, damaged or defective clothing directly from The North Face. In the future, consumers will also be able to sell their own worn The North Face gear.

The importance of repair has also been identified by policymakers. In March 2023, the European Parliament and Council reached an agreement on common rules to promote the repair of goods for consumers. Once adopted, the new rules will introduce a new 'right to repair' for consumers, both within and beyond the legal guarantee, which will make it easier and more cost-effective for them to repair products instead of simply replacing them with new ones. To boost the development of the repair market, the new rules will ensure that spare parts for technically repairable goods are made available at a reasonable price, and that manufacturers will be prohibited from using contractual, hardware or software related barriers to repair [18].

5.1.7 Product-Service Systems (PSS)

Business models based on product-service systems (PSS) establish a value proposition focused on final users' needs rather than on the product, allowing for an easier design of a need-fulfilment system, which in turn offers environmental and social benefits [18]. According to Tukker, PSS can be classified into three different main categories [20]. These categories are summarized in Table 5.7:

Table 5.7 Main Categories of Product-Service Systems [20]

PSS Category	Characteristic
Product-oriented services	The business model is still mainly geared toward sales of products, but some extra services are added
Use-oriented services	The traditional product still plays a central role, but the business model is not geared toward selling products. The product stays in ownership with the provider, and is made available in a different form, and sometimes shared by a number of users
Result-oriented services	The client and provider in principle agree on a result, and there is no pre-determined product involved

Examples offor use-oriented PSS in the textile industry are provided in Table 5.8:

Table 5.8 Examples of Use-Oriented PSS in the Textile Industry [21]

Example	Characteristic
Product lease	Provider maintains ownerships of clothing, and consumers pay a membership or subscription fee for access to a wardrobe or garments for a certain amount of time, typically a month. Garments are sequentially used by different users.
Product renting or sharing	Provider maintains ownerships of clothing, and consumers pay-per-use for garments, typically very short term of 2 to 14 days. Garments are sequentially used by different users.
Product pooling	Typically seen as a clothing library, where clothing ownership can either be with the provider/organizer or remain with the garment owner. Typically, consumers pay a membership fee in which there is unlimited access to available garments in the library, typically for a month or a designated amount of time.

Providing a product as a service can change consumption patterns and may provide incentives for the optimization of supply chains and product design to maximize the value offering to the consumer and the company. This could result in extended life-spans of products, higher use intensities, and other value chain optimizations. However, the real-world impact of PSS on reducing consumption is uncertain. Strategies to extend product longevity for clothing are partly compromised by external factors such as changes in style and fashion obsolescence, which shorten product longevity regardless of the material durability. Generally, the effectiveness of PSS depends on significantly replacing primary production, which is often not the case in practice. Some consumers may choose rental or sharing initiatives to increase their wardrobe choice rather than replace their usual purchases. In this context, it is important to consider that some business models might even promote consumption, for instance by offering regular replacement of a product. Consequently, true sustainability requires not only innovative business models but also changes in consumer habits and behaviors to ensure substantial environmental benefits [21].

References for Section 5.1

[1] Veit, D., *Fibers* (2023), Springer, Charm

[2] N.N., *Textile Recycling Market Size, Share Trends Analysis Report By Material (Cotton, Polyester, Wool, Polyamide), By Source (Apparel Waste), By Process (Mechanical), By Region, And Segment Forecasts, 2023–2030 (2023)*, *https://www.grandviewresearch.com/industry-analysis/textile-recycling-market-report*. Last accessed 10 July 2024

[3] N.N., *Kreislaufwirtschaft: Definition und Vorteile (2023)*, *https://www.europarl.europa.eu/topics/de/article/20151201STO05603/kreislaufwirtschaft-definition-und-vorteile*. Last accessed 10 July 2024

[4] Gino Kraft, M.H., Christ, O., Scherer, L., *Management der Kreislaufwirtschaft* (2022), Springer Gabler, Wiesbaden

[5] N.N., *https://dictionary.cambridge.org/dictionary/english/greenwashing*. Last accessed 10 July 2024

[6] Cobing, M., Wohlgemuth, V., Vicaire, Y., Greenwash Danger Zone (2023), *https://www.greenpeace.de/publikationen/Greenpeace_Report_Greenwash_Danger_Zone.pdf*. Last accessed 10 July 2024.

[7] Geissdoerfer, M., Pieroni, M.P.P., Pigosso, D.C.A., Soufani, K., *J. Clean. Prod.* (2020) 277, 123741; DOI: *10.1016/j.jclepro.2020.123741*

[8] Letmathe, P., *Strategies for Transforming to a Circular Economy: New Business Models*. Lecture as part of the 6th ITMF-ITA Webinar: Circular textile Economy (2024)

[9] Trenz, E., *Textilplus* (2021) 01/02, pp. 14–16

[10] Gries, T., Veit, D., Wulfhorst, B., In *Textile Fertigungsverfahren*. Greis, T., Veit D., Wulfhorst, B. (Eds.) (2009), Hanser, Munich

[11] Arafat, Y., Uddin, A.J., *Heliyon* (2022) 8, e11275; DOI: *10.1016/j.heliyon.2022.e11275*

[12] Ribul, M., Lanot, A., Tommencioni Pisapia, C., Purnell, P., McQueen-Mason, S., Baurley, S., *J. Clean. Prod.* (2021) 326, 129325

[13] N.N., *https://www.rittec.eu/*. Last accessed 10 July 2024

[14] Pires, A., Martinho, G., Rodrigues, S., Gomes, M.I., Sustainable Solid Waste Collection and Management (2019), Springer, Charm

[15] Aus, R., Moora, H., Vihma, M., Unt, R., Kiisa, M., Kapur, S., *Fash Text* (2021) 8, 34; DOI: *10.1186/s40691-021-00262-9*

[16] Prieto-Sandowal, V., Jaca, C., Ormazabal, M., *J. Clean. Prod.* (2018) 197, pp. 605–615; DOI: *10.1016/j.jclepro.2017.12.224*

[17] N.N., *https://dictionary.cambridge.org/dictionary/english/repair*. Last accessed 10 July 2024

[18] N.N., *Commission welcomes political agreement on new consumer rights for easy and attractive repairs* (2024), *https://ec.europa.eu/commission/presscorner/detail/en/ip_24_608*. Last accessed 10 July 2024

[19] Annarelli, A., Battistella, C., Costantino, F., Di Gravio, G., Nonino, F., Patriarca, R., *CIRP-JMST* (2021), 32, pp. 424–436; DOI: 10.1016/j.cirpj.2021.01.010

[20] Tukker, A., *Bus. Strat. Env.* (2004) 13, pp. 246–240; DOI: 10.1002/bse.414

[21] Johnson, E., Plepys, A., *Sustainability* (2021) 13, 2118; *10.3390/su13042118*

5.2 Fashion Design for Recycling: Possible Eco-Design Strategies for the Success of Progressive Sustainability Solutions in the Fashion and Textile Industry

Chiara Colombi, Erminia D'Itria, Design Department, Politecnico di Milano, Italy

5.2.1 Introduction

The fashion industry, known for its constant innovation and dynamism, has witnessed a transformative shift in recent years. In the past, fashion companies prioritized the extraction of natural resources and waste disposal according to the linear development model of mass production and consumption. Such a model promoted excessive use and early disposal of garments and focused on fulfilling desires rather than needs [10]. However, the introduction of disruptive eco-design innovations in materials and manufacturing practices is revolutionizing the industry. By adopting an ecological perspective, the fashion industry is now striving to achieve higher levels of sustainability, a stark departure from the linear model, which has historically contributed to resource depletion and environmental degradation.

In this scenario, textile and fashion companies, often small and medium-sized enterprises (SMEs), have been at the forefront of adopting ecological approaches based on scientific knowledge, driven by designing for resource-efficient manufacturing [14]. They are reimagining traditional production methods and experimenting with innovative techniques to enable their resources' production, recovery, and transformation. These eco-design approaches aim to decouple fashion innovation from resource exploitation by emphasizing alternative materials and recycling practices [11].

Selected case studies illustrate how, from a design perspective, adopting eco-design principles in the fashion industry is multi-faceted. This includes developing and using sustainable materials, such as recycled and organic fabrics, and exploring innovative and environmentally friendly manufacturing processes. These processes encompass everything from reducing manufacturing water consumption and pollution to implementing practices that encourage consumers to repair and recycle their clothing. By embracing new pathways rooted in eco-design principles, fashion companies are redefining their routines, reducing resource consumption and minimizing their ecological footprint. They are pioneering innovative approaches to resource-efficient manufacturing, ultimately working toward a circular economy where fashion items are designed, produced, used and recycled sustainably and in an environmentally responsible manner. This shift is crucial for the long-term health of the industry and the well-being of the planet and its inhabitants. The fashion industry's focus on eco-design, therefore, not only reduces its environmental footprint, but also enhances its social responsibility by promoting fair labor practices and supporting local communities [12].

5.2.2 Codifying Eco-Design Approaches in Fashion

The fashion industry stands globally as one of the largest industrial sectors, featuring a diverse and intricate production system that spans every stage of the supply chain. Its companies struggle with the challenges posed by this production system and its economic models, which create and encourage an unsustainable pattern of fashion consumption [3]. As Gwilt Rissanen (2012) discussed, the fashion industry is defined by speed and uncertainty, and a focus on producing fashionable products that foster the idea of planned obsolescence. This approach introduces new clothing at low prices and with great speed, affecting the end product's overall quality. When it comes to creating, using and disposing of fashion items, the sheer volumes fueled by mass consumption have a significant environmental impact. In discussing this ecological impact, it is crucial to understand that the term "environment" encompasses various contexts, including economic, social and environmental aspects. In this broader context, fashion is pivotal and intricately connected to the overall economic, social and environmental ecosystem [16]. Accordingly, it is imperative to incorporate environmental considerations into the fashion product development process with a view to balancing ecological concerns and socio-economic requirements.

In this context, eco-design is one of the main drivers for shifting the existing system toward effectively adopting new sustainable development models [23]. Eco-design integrates environmental aspects at every stage of product development, aiming to create products with minimal environmental impact throughout their life cycle [7]. Such a perspective enables a product's environmental impacts at various stages of the life cycle to be assessed and those which have the most significant effects to be pinpointed. Consequently, it provides valuable insights for strategic design interventions through a multi-criteria approach that considers all the conventional product requirements and, at the same time, combines them with pertinent environmental factors and their associated impacts [4]. To guide designers in such strategic processes and support the development of products with enhanced environmental performance, several scholars have proposed eco-design principles (EDPs) based on different environmental perspectives [4, 15, 28, 30].

The authors have revisited and reorganized the principles proposed by Maccioni et al. (2019) into ten core environmental considerations, grouping the initial set of 66 principles on the basis of their conceptual nuances and categorizing them according to life-cycle stage, from the design phase to the end of life (EoL) management.

From an eco-design perspective, these principles address three main strategic actions that design can implement to foster a sustainable transformation: (1) maximize the use of sustainable materials, (2) use the least amount of energy required and (3) design products to be reusable at the end of their life cycle (Table 5.9).

Table 5.9 List of Eco-Design Principles

Principle	Description	Life-Cycle Stage	Maximize Use	Use Less	Be Reusable (EOL)
P1	Adoption of materials with less environmental impact	Creation and design		X	X
P2	Use of fewer materials in the manufacture of products	Creation and design Manufacturing	X	X	
P3	Utilization of fewer resources during manufacturing (water, energy, air)	Manufacturing	X	X	
P4	Production of less pollution and waste	Manufacturing		X	X
P5	Decreasing the environmental impact of product distribution	Distribution	X		
P6	Design of products that use fewer resources when used by end customers	Creation and design Use	X	X	X
P7	Design of products that cause less waste and pollution during use	Creation and design Use		X	X
P8	Optimizing the functionality of products to ensure they last longer	Creation and design Use	X		
P9	Enabling reuse and recycling	Creation and design EoL management			X
P10	Reducing the environmental impact of disposal	Creation and design EoL management	X		X

The adoption of these principles in fashion design and production processes can significantly reduce the challenges raised by the fashion industry's resource-intensive and linear production processes. These eco-design principles offer a structured framework for addressing these challenges and transitioning toward a more sustainable and environmentally responsible fashion industry.

One aspect that emerges from these principles is how design plays a crucial role in the envisioning and implementing of production systems that maximize the use of already-processed resources, avoid the exploiting of new, virgin and nonrenewable resources and are oriented toward continuous resource reuse.

The authors conducted a comprehensive analysis of various design methodologies encompassed by the broad concept of "Design for X." This term includes a range of sustainable design strategies, such as designing for recycling, enhancing recyclability, facilitating ease of dismantling, and improving repairability [7, 8]. By systematically comparing these approaches, the study aimed to identify which design principles most effectively align with the core objectives of eco-design. These principles were then grouped to establish a coherent framework for sustainable design practices within the fashion and textile industries.

Design for recycling is the approach embracing all these aspects to move fashion toward a sustainable paradigm. By strongly emphasizing recyclability and circularity throughout the design process, this approach aims to minimize waste and promote the responsible utilization of resources. It encourages the selection of materials that can be readily repurposed, ultimately contributing to the establishment of a more circular economy with an efficient use of resources and minimum waste. This represents a substantial step toward reducing the fashion industry's environmental footprint and fostering a more environmentally aware approach to fashion production and consumption.

The selected case studies presented in the following section serve to investigate how fashion companies address the numerous environmental challenges posed by their resource-intensive and linear production processes from the specific perspective of design for recycling.

5.2.3 Fashion Design for Recycling

Fashion design for recycling (FDfR) develops strategies to design and redesign garments and textiles, ensuring that each item exists in the highest-possible-value form for as long as possible. Such an approach embeds a systemic call to encourage textile companies, brands, and designers to work on their waste (stock, prototypes, etc.) and to incorporate it into the design of collections and capsules through aesthetics, quality, price, and market positioning [26]. Furthermore, such an approach also stimulates partnerships, ranging from manufacturers to local communities to companies that commit to emphasizing their responsible practices to collectively rethink and reimagine the garment life cycle [29].

To achieve these goals, companies engage in collection, sorting and processing. The collection phase involves retrieving products discarded by consumers [27]. The sorting phase selects discarded products by condition or type [20]. Finally, the processing phase focuses on design-led actions for restoring the function of products or materials and increasing their value [1]. Despite the benefits of recycling, companies operating in this dimension face several challenges. The main issues are (1) inadequate infrastructure for handling material waste, (2) regulations crafted for linear systems and (3) the misconceptions among consumers about the durability of products and second-hand materials present obstacles that demand innovative solutions [21].

However, these challenges have not discouraged small and medium-sized enterprises (SMEs) from investing in design-oriented approaches that diminish reliance on natural resources and bolster the reutilization of fashion items [11]. Fashion and textile companies are harnessing the potential of eco-design knowledge to make the hidden value within recycled products explicit and enable their recirculation toward a circular and sustainable productive model [5].

In the context of growing environmental awareness and the urgent need for circularity in fashion, this section presents specific design trajectories that support recycling-oriented fashion design. These trajectories represent strategic pathways that designers can follow to develop sustainable fashion and textile solutions. They reflect the industry's shift toward circular, resource-efficient systems and offer concrete examples or approaches that can inspire and inform future design practices. The main goal is to explore how different stages of the supply chain can be reconnected through the reuse and transformation of materials, fostering new synergies across the system.

The embracing of fashion design for recycling and the adoption of progressive eco-design approaches are pivotal steps toward mitigating the environmental impact of the fashion and textile industry while fostering innovation and creativity. Recycling practices not only reduce the industry's dependence on natural resources but also lead to the development of innovative and ecological products. As more companies and consumers adopt eco-design principles, the industry can move closer to a future where waste is minimized, resources are conserved, and fashion becomes genuinely sustainable.

In the context of growing environmental awareness and the urgent need for circularity in fashion, this section introduces identified design trajectories aimed at supporting fashion design for recycling. This scenario reflects the current shift in the industry toward more circular, resource-efficient systems. These trajectories guide design directions that serve as inspiration for framing sustainable fashion and textile systems. The primary objective is to explore potential pathways for creating new synergies along the supply chain, by reconnecting different stages through the reuse and transformation of processed materials. These initiatives focus on eliminating waste while fostering local ecosystems and laying the foundation for a closed-loop supply chain. This innovative approach disrupts conventional practices by strategically intervening at both the upstream phase (during the design process) and the downstream step (in waste management) of the supply chain. The goal is to generate a cascade effect, influencing subsequent stages of the process (Figure 5.3).

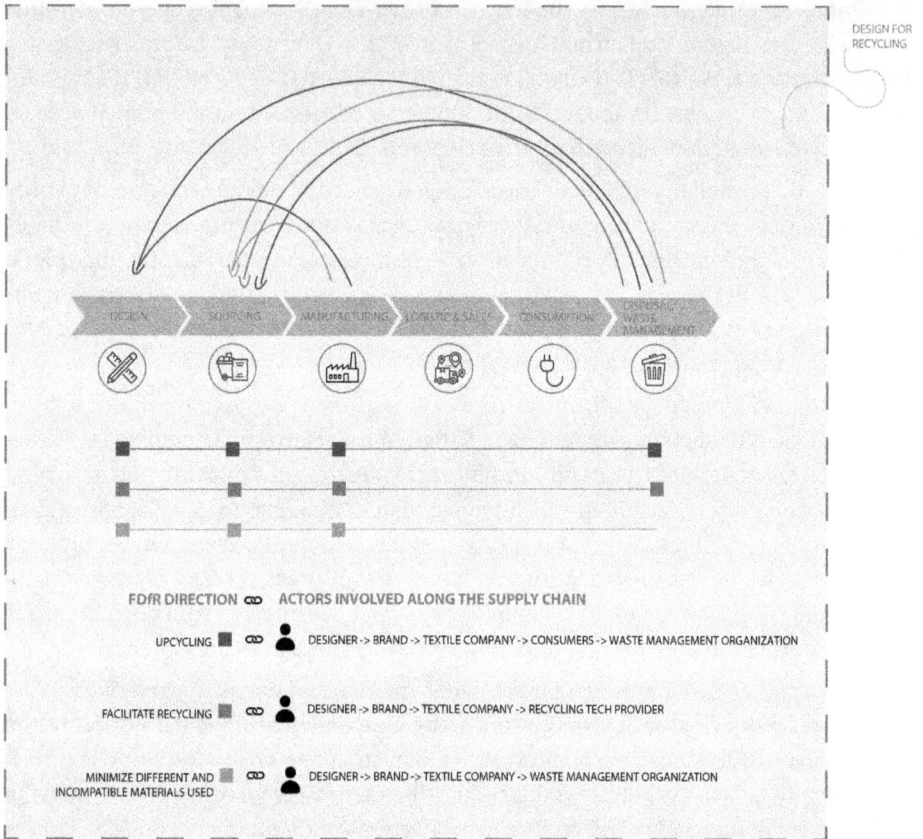

Figure 5.3 The interpretative model of the fashion design for recycling (FDfR) system

From an operational design perspective, this study identifies three key thematic pathways that guide designers in implementing recycling practices within the fashion industry:

- Upcycling
- Designing to facilitate recycling
- Minimizing the use of incompatible materials

These pathways emerge from the analysis of current sustainable design strategies and offer practical directions for embedding circularity into fashion and textile systems.

The first pathway, named upcycling, has emerged as a creative and impactful strategy for enhancing sustainability in fashion. Unlike traditional recycling or downcycling, which often reduces the quality of materials, upcycling adds value to discarded or surplus textiles by transforming them into new, high-quality products. This approach re-imagines textile waste – such as deadstock, offcuts, and pre-owned garments – diverting it from landfills and incineration. It fosters innovation, craftsmanship, and collaboration among designers, brands, and consumers. By encouraging circular thinking and revalorizing resources, upcycling challenges the linear production model and aligns with the principles of ecological responsibility and conscious consumption.

In parallel, the second pathway focuses on designing garments to facilitate recycling by addressing product recyclability from the outset. Many garments today are composed of fiber blends that are difficult to separate, making them unsuitable for recycling. Designers play a crucial role in overcoming this barrier by selecting recyclable materials and creating products that are easy to disassemble at the end of their life cycle. Technological innovations now support this effort, offering solutions such as modular construction and automated separation systems. This approach enhances collaboration between designers, manufacturers, and recycling technology providers, enabling the creation of garments that are not only functional and aesthetically appealing but also prepared for responsible end-of-life processing.

The third pathway emphasizes minimizing the use of incompatible or mixed materials, which remains a major obstacle to effective recycling. Simplifying material composition and ensuring compatibility are essential to make recycling feasible. While advancements in chemical and mechanical recycling technologies are improving the ability to break down complex materials, the application of eco-design principles from the earliest design stages remains fundamental. By choosing compatible materials and construction methods that enable easy disassembly and transformation, designers can ensure that garments are more easily recycled or repurposed. This contributes to the broader shift toward circularity and requires coordinated action across the supply chain to support the sustainable management of fashion products throughout their entire life cycle.

Collectively, these three pathways provide a strategic framework for integrating recycling principles into fashion design. They demonstrate how thoughtful material choices and design approaches can advance circular economy objectives, reduce environmental impact, and drive systemic transformation toward a more sustainable textile industry.

To illustrate these concepts in practice, three case studies follow – each aligned with one of the proposed trajectories. These examples are emblematic of the thematic directions discussed and reflect how eco-design principles are being applied in real-world contexts. The selected companies and designers are distinguished by their contributions to sustainable fashion, their commitment to circular models, and their role in shaping a new mindset regarding the responsible use and consumption of the planet's finite resources.

5.2.3.1 Upcycle: Christopher Raeburn

Christopher Raeburn pioneered the upcycling movement and established his eponymous brand in 2008. Raeburn is known for his innovative use of British Army parachutes as his primary fabric source for creating new garments. Instead of starting with traditional textiles, he saw the potential in these surplus military materials and found ways to turn them into fashionable pieces. Raeburn's expertise extends to reconstructing various military surplus items, including parachutes, parkas and military jackets. He refers to this process as "reappropriation," a term that emphasizes the reutilization of these materials. What sets Raeburn apart is the choice of materials and his unique approach to design. He bases his design decisions on thorough research of the source materials and the deconstruction process itself. Therefore, the characteristics of the materials influence the shape, silhouette, and overall design of his garments. Instead of imposing a preconceived design onto the fabric, he lets the materials guide and inform the final product. This fashion approach contributes to sustainability by reducing waste and adds a distinctive, creative element to the design process.

5.2.3.2 Facilitate Recycling: Resortecs

Resortecs is an innovative technology company that aims to achieve complete circularity within the textile and fashion industries. It is developing and implementing innovative technologies and methods to make recycling a fast and practical process for fashion brands, recyclers, and all participants in the supply chain. Specifically, the company has developed a unique dissolvable stitching thread and heat-dismantling technology for garments. This technology allows for easier disassembly of clothing at the end of its life cycle, which is a critical step in promoting recycling and upcycling within the fashion industry. The dissolvable stitching thread withstands a garment's everyday wear and tear while completely dissolving when exposed to a specific heat level. This enables the easy separation of different textile components, such as zippers, buttons and other fasteners, and facilitates the recycling and reuse of materials. Resortecs' approach considers three elements: technology, method, and circular commitment. The company employs cutting-edge technology to develop garments and textiles that can be easily disassembled and recycled. This technology involves specialized stitching techniques that facilitate the separation of components and the recovery of valuable materials. Its method guides stakeholders to its systematic and practical solution. This includes guidelines, best practices and tools for designing textiles and clothing with recycling in mind. Its commitment to full circularity aims to achieve a closed loop, which means that materials from old garments are efficiently reused to create new ones. This commitment reflects a broader sustainability goal of reducing waste and environmental impact. Resortecs represents a much-needed solution in the fashion industry that provides a clear path toward a more sustainable and environmentally friendly future. As the world increasingly recognizes the importance of reducing waste and adopting circular practices, Resortecs' work is a pioneering example of how technology and innovation can lead the way to a more responsible and ethical fashion industry.

5.2.3.3 Minimize Different or Incompatible Materials: Napapijri

Napapijri is an Italian apparel brand that has implemented a forward-thinking approach to sustainability by simplifying the design of its jackets in its Circular Series. In this series, the entire coat, including the fabric, filling, and trimmings, is exclusively made from one material: nylon 6. This design choice is significant for several reasons. Nylon 6 is a type of nylon that can be recycled efficiently. The use of a single material throughout the jacket streamlines the recycling process and makes it easier to recover and repurpose the fabric when the coat ends its useful life. To ensure the activation of this recycling process, Napapijri encourages customers who purchase a Circular Series jacket to register their items online. This registration is a crucial step in the product's life-cycle management because it allows tracking and managing the coat when recycling is due. The brand has partnered with Aquafil, its nylon supplier, to give back the worn-out jackets and guarantee a closed-loop chemical recycling process. This process breaks down the old jackets into constituent materials, including nylon 6. Aquafil creates Econyl®, a recycled nylon 6 yarn from this recycled material. The regenerated nylon from the old jackets is then ready to manufacture new Circular Series Jackets. The material from old jackets is reincorporated into new ones, promoting a circular, sustainable approach to fashion. Napapijri's Circular Series not only represents an innovative and eco-conscious approach to fashion design, but also highlights the importance of taking responsibility for the entire life cycle of a product, from creation to recycling and reuse. It is a model that demonstrates how sustainable practices can be integrated into the fashion industry, thereby reducing waste and promoting the use of recycled materials.

5.2.4 Conclusions

In conclusion, the fashion industry is undergoing a transformative shift by significantly emphasizing alternative materials and recycling practices. The presented approaches proposed a change from the traditional model of resource exploitation, as they not only reduce the industry's dependence on finite resources, but also provide a fundamental shift in the creation, consumption and disposal of fashion. One of the fundamental driving forces behind this transformation is the adoption of eco-design principles. From the utilization of upcycled materials to the development of innovative recycling techniques, these practices are proving to be catalysts for change within the industry. They contribute to the establishment of a more circular system where secondary raw materials are collected, reprocessed and recycled throughout the supply chain. This shift represents a significant departure from the linear, wasteful model that has characterized the fashion industry for decades and it moves us closer to a more regenerative approach. Case studies showcased in this chapter provide tangible evidence of how fashion brands actively pursue this objective. Examples range from collecting used garments for recycling to implementing closed-loop production processes, all contributing to a more sustainable and environmentally conscious fashion industry.

Nevertheless, the journey toward a fully sustainable and circular fashion industry has challenges. While recycling practices hold great potential, they require substantial research, development, and infrastructure investments. Collaboration among designers, manufacturers, and consumers is pivotal in ensuring the broader adoption of eco-design principles and achieving these sustainability goals. As the fashion industry grapples with its ecological impact, the growing awareness of these issues is sparking a powerful paradigm shift. With global demand for apparel and accessories at an all-time high, the industry must confront the environmental consequences of its rapid production and disposal processes. The integration of eco-design principles into the fashion world represents a significant step forward, bringing us closer to a future where the industry minimizes its environmental footprint and deeply embeds sustainability in every facet of its operations.

References for Section 5.2

[1]   D'Itria, E., Vacca, F. (2023). How can social-cultural values nurture sustainability in the Fashion sector?. In Global Fashion Conference 2022 (pp. 1–14).

[2]   Elf, P., Werner, A., Black, S. (2022). Advancing the circular economy through dynamic capabilities and extended customer engagement: Insights from small sustainable fashion enterprises in the UK. Business Strategy and the Environment, 31(6), 2682–2699.

[3]   D'Itria, E., Aus, R. (2023). Circular fashion: evolving practices in a changing industry. Sustainability: Science, Practice and Policy, 19(1), 2220592.

[4]   da Silva, F.M. (2018). Sustainable fashion design: Social responsibility and cross-pollination. Textiles, Identity and Innovation: Design the Future, 439–444.

[5]   Bocken, N.M., Short, S.W., Rana, P., Evans, S. (2014). A literature and practice review to develop sustainable business model archetypes. Journal of Cleaner Production, 65, 42–56.

[6]   Gwilt, A., Rissanen, T. (Eds.). (2012). Shaping sustainable fashion: Changing the way we make and use clothes. Routledge.

[7]   Fletcher, K. (2010). Slow fashion: An invitation for systems change. Fashion Practice, 2(2), 259–265.

[8]   Kim, H., Cluzel, F., Leroy, Y., Yannou, B., Yannou-Le Bris, G. (2020). Research perspectives in eco-design. Design Science, 6, e7.

[9]   Ceschin, F., Gaziulusoy, İ. (2019). Design for Sustainability (Open Access): A Multi-level Framework from Products to Socio-technical Systems. Routledge.

[10] Bovea, M.D., Pérez-Belis, V. (2012). A taxonomy of eco-design tools for integrating environmental requirements into the product design process. Journal of Cleaner Production, 20(1), 61–71.

[11] Vezzoli, C., Manzini, E. (2017). Design for sustainable consumption and production systems. In System Innovation for Sustainability 1 (pp. 148–168). Routledge.

[12] Telenko, C., O'Rourke, J.M., Conner Seepersad, C., Webber, M.E. (2016). A compilation of design for environment guidelines. Journal of Mechanical Design, 138(3), 031102.

[13] Fiksel, J. (2009). Design for environment: a guide to sustainable product development. McGraw-Hill Education.

[14] Maccioni, L., Borgianni, Y., Pigosso, D.C. (2019). Can the choice of eco-design principles affect products' success? Design Science, 5, e25.

[15] Chiu, M.C., Kremer, G.E.O. (2011). Investigation of the applicability of Design for X tools during design concept evolution: a literature review. International Journal of Product Development, 13(2), 132–167.

[16] Niinimäki, K., Karell, E. (2020). Closing the loop: Intentional fashion design defined by recycling technologies. Technology-Driven Sustainability: Innovation in the Fashion Supply Chain, 7–25.

[17] Todeschini, B.V., Cortimiglia, M.N., de Medeiros, J.F. (2020). Collaboration practices in the fashion industry: Environmentally sustainable innovations in the value chain. Environmental Science Policy, 106, 1–11.

[18] Nørup, N., Pihl, K., Damgaard, A., Scheutz, C. (2019). Evaluation of a European textile sorting center: Material flow analysis and life cycle inventory. Resources, Conservation and Recycling, 143, 310–319.

[19] Hawley, J.M. (2006). Textile recycling: A systems perspective. In recycling in textiles. Woodhead Publishing Limited, UK.

[20] Abraham, N. (2011). The apparel aftermarket in India–a case study focusing on reverse logistics. Journal of Fashion Marketing and Management: An International Journal, 15(2), 211–227.

[21] Juanga-Labayen, J.P., Labayen, I.V., Yuan, Q. (2022). A review on textile recycling practices and challenges. Textiles, 2(1), 174–188.

[22] Cassidy, T.D., Han, S.L.C. (2017). Upcycling fashion for mass production. In sustainability in fashion and textiles (pp. 148–163). Routledge.

[23] Cassidy, T.D., Han, S.L. (2013). Upcycling fashion for mass production. In A.L. Torres, M.A. Gardetti (Eds.), Sustainable Fashion Textiles: Values, design, production and consumption (pp. 148–163). Routledge, Taylor Francis Group.

[24] Aus, R., Moora, H., Vihma, M., Unt, R., Kiisa, M., Kapur, S. (2021). Designing for circular fashion: integrating upcycling into conventional garment manufacturing processes. Fashion and Textiles, 8, 1–18.

[25] Ekström, K.M., Salomonson, N. (2014). Reuse and recycling of clothing and textiles – A network approach. Journal of Macromarketing, 34(3), 383–399.

[26] Karell, E., Niinimäki, K. (2019). Addressing the dialogue between design, sorting and recycling in a circular economy. The Design Journal, 22(sup1), 997–1013.

[27] Colombi, C., D'Itria, E. (2023). Fashion Digital Transformation: Innovating Business Models toward Circular Economy and Sustainability. Sustainability, 15(6), 4942.

[28] Goldsworthy, K., Earley, R., Politowicz, K. (2018). Circular speeds: a review of fast slow sustainable design approaches for fashion textile applications. Journal of Textile Design Research and Practice, 6(1), 42–65.

[29] Harmsen, P., Scheffer, M., Bos, H. (2021). Textiles for circular fashion: The logic behind recycling options. Sustainability, 13(17), 9714.

[30] Medkova, K., Fifield, B. (2016). Circular design-design for circular economy. Lahti Cleantech Annual Review, 32.

6
Circularity of Technical Textiles and Composites

6.1 Circularity of Technical Textiles

Vanessa Overhage, Institut für Textiltechnik of RWTH Aachen University, Germany

Technical textiles are all textiles that are primarily used for their technical properties. The applications range from textiles to improve acoustics to protective clothing to nonwovens for various filter applications to fabrics as conveyor belts in production and reinforcement materials in the construction industry. The areas of application are diverse. Depending on the application, the desired properties are achieved through the materials, combinations of materials or processing of them into textile surface structures, and post-treatment, such as coatings or a combination of these.

6.1.1 Categories of Technical Textiles

Technical textiles are divided into 12 categories based on their application sector. The categories were defined at the "Techtextil" international trade fair in 2009. Figure 6.1 presents an overview of the different categories.

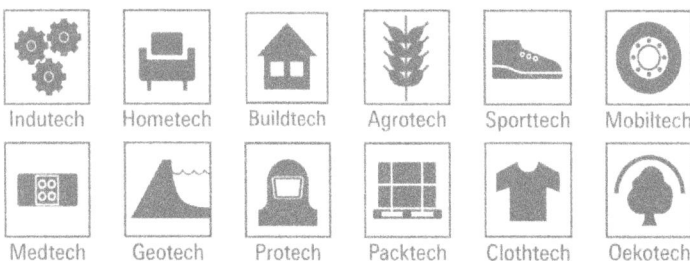

Figure 6.1 Categories of technical textiles based on Techtexil [1]; [Pictrograms: © Techtextil, Messe Frankfurt Exhibition GmbH, Frankfurt am Main]

Indutech covers all technical textiles that are used for industrial purposes, for example, in mechanical engineering and the chemical and electrical industries. These include conveyor belts, belts and filters. **Hometech** comprises home and household textiles. This category includes floor coverings, room dividers, indoor and outdoor seating furniture and insulation materials. Among other things, the technical textile solutions influence sound absorption and thus room acoustics, the resistance of the materials to external influences, or integrate light. **Buildtech** consists of all technical textiles for building with membranes and in lightweight and solid construction. Glass and carbon textiles are used, for example, for the reinforcement of concrete components. This combination of materials is known as textile-reinforced concrete (see Figure 6.2 (b)). Heat and fire-resistant materials are used, for example, to protect against fires and in hazardous areas. On the other hand, membranes and tent roofs are made of waterproof coated textiles. Other applications for textiles include privacy screens and curtain walling. **Agrotech** covers all technical textiles used in agriculture and forestry. In addition to fishing nets, textiles serve as thermal protection, to prevent water evaporation and to protect plants from damage, due to environmental influences. Another area of application is in the field of gardening and landscaping, where technical textiles are used as artificial turf, as a protection against frost and as planting mats. Old coffee bags made from jute fibers are, for example, repurposed as frost protection or recycled and produced as a nonwoven that can serve as a planting mat. **Sporttech** covers sports equipment made of fiber-composite materials or technical textiles in the form of stunt kites, sails and heatable clothing. Bicycle frames, sailing boats and surfboards are made from fiber-composite materials. Smart textiles are used for lightening in clothing for outdoor sports. Knitting and woven structures for shoes are also part of Sporttech.

Mobiltech comprises components of cars, trains and the aerospace industry, among others. In addition to structural components, tire reinforcements and interior paneling are also classified as Mobiltech. Seat belts and airbags can also be assigned to this category, but also to Protech. Furthermore, Mobiltech includes seat upholstery for cars, planes and trains with a spacer fabric for improved air circulation. **Medtech** includes textiles and hygiene products for the medical sector: diapers, wound dressings and surgical clothing. In addition, implants and explants are also a great part of technical textiles in Medtech, including stents and bandages. Furthermore, technical textiles combined with smart textiles can serve as a movement sensor in a carpet for the detection of falls and strokes. **Geotech** covers geotextiles for erosion protection, dyke and slope stabilization (see Figure 6.2 (a)) as well as reinforcement textiles for road and infrastructure construction. Technical textiles are used on landfill sites and in wastewater treatment plants to protect or rather filter the groundwater. **Protech** not only includes the aforementioned airbags and safety belts, but also protective clothing for various fields of application. Depending on the desired function, the protective textiles are designed with heat-resistant, temperature-regulating or cut-resistant materials. Use-cases include fire fighters, police officers, astronauts and craftsmen. Furthermore, water-repellent umbrellas and fire blankets rank among the textiles in this category. **Packtech** contains textiles used as packaging material, such as jute bags for shopping, as well as packaging nets for food or big bags for bulk goods in industry and on construction sites. Tear-resistant lashing straps, safety ropes and lightweight transport trolleys in airplanes are based on technical textiles. **Clothtech** includes clothing textiles that serve as protective equipment, but also smart textiles with an integrated light or headphones for sports and leisurewear. Wearable airbags and heatable winter jackets are examples of Clothtech. **Oekotech** comprises technical textiles for environmental protection and recycling. Textiles for erosion protection, filter textiles and protective nets for fruit trees are also assigned to this category. Furthermore, recycled materials find application as yarns made out of cork or from PET bottles to produce clothing or technical textiles [1].

Figure 6.2 Technical textiles from different categories (Geotech (a), Buildtech (b), Oekotech (c), Hometech (d))

6.1.2 The Market for Technical Textiles

The estimated size of the global market for technical textiles was 172 billion euros in 2022 and is expected to reach 180 billion euros in 2023. Continued growth is projected and is estimated to be worth 248 billion euros in 2030 [2].

Asia-Pacific dominated the global technical textiles market by both value and volume in 2018. The market is expected to continue growing, due to the mass availability of raw materials and the favorable growth of end-use industries, such as automotive, construction, packaging and apparel. India is projected to be the fastest-growing market for technical textiles. The Indian government's increasing investments to support the textile industry and rapid industrialization are creating potential growth opportunities for the market there. According to market research studies, North America is the second-largest consumer, end-user, and importer of technical textiles, and the demand for these textiles is expected to grow further. The study shows that Europe holds the third-largest share of the global technical textiles market, accounting for 18% of consumption. Germany has the highest sales share within Europe. The demand for technical textiles is expected to increase in Germany, due to the growth of the automotive industry, rising safety standards, advancements in the medical sector, and the expansion of wind-turbine rotor blades. This trend is expected to be reflected worldwide [3].

6.1.3 Circular Use of Technical Textiles

The circular economy is becoming increasingly relevant for technical textiles. It involves the use of textiles beyond the end of their first life cycle. Given the scarcity of resources, increasing demand, and sustainability concerns, it is important to consider the "R-strategies" of refuse, rethink, reduce, reuse, repair, refurbish, remanufacture, repurpose and recycle [4]. The following section presents examples of different R-strategies for the aforementioned categories of technical textiles, e.g. the reduction of plastic-based geotextiles and the use of biodegradable natural fiber based geotextiles, the scope for repurposing wind-turbine blades and the use of recycled fibers in the Buildtech Industry.

6.1.3.1 Reducing Plastic-Based Textiles in Geotech – Biodegradable Technical Textiles

Geotextiles are used for erosion protection, dyke and slope stabilization as well as reinforcement textiles for road and infrastructure construction. They are supposed to have a long lifetime, for example, when used in infrastructure construction. But the necessary lifetime of geotextiles is in some cases limited by their use-case. It is therefore possible to use materials that degrade within a certain time frame in the environment. Such an instance would be technical textiles used for dyke stabilization or for the stabilization of erosion-prone slopes, where plants are intended to grow and fix the soil with their roots over time. After a certain period of time, it is desired that the textile degrades so that they do not to be unnecessarily removed because of the plant roots (see Figure 6.3).

Figure 6.3 Natural-fiber-based technical textiles – biodegradable over time ((a) new geotextile, (b) already partially degraded geotextile)

6.1.3.2 Repurposing – A Second Life with a New Purpose

Technical textiles have certain property requirements. To ensure that these are met, some products have a fixed lifetime for safety reasons. For example, wind turbines are currently designed for a service life of around 20 years. Due to weather conditions and exposure constant vibrations from the wind, their lifetime is limited. Another example is seat belts in cars, which also have a limited lifetime, due to the end-of-life of the car or if the seat belt gets damaged beforehand.

Neither wind turbines nor seat belts will be reused for the same purpose. In addition to recycling approaches, there are various ideas for repurposing. Designers are creating items such as handbags, backpack fasteners and shoulder straps for bags made from seat belts. Understandably, the applications for discarded wind-turbine rotor blades are different; these are being redesigned for constructions on playgrounds, for pedestrian bridges, bicycle shelters and other components, whereby the basic structure of the rotor blades remains visible.

6.1.3.3 Recycling – The Use of Recycled Carbon Fibers in Buildtech

The construction sector is the most resource-intensive industry. Concrete is currently the most widely used building material globally [5]. It consumes a significant amount of resources, with an annual requirement of 30 billion tons [6]. The primary constituents of concrete are water, cement, and aggregate, such as gravel and sand. All of these materials are non-renewable resources. Sand is currently the most important trading raw material worldwide and competes with a wide variety of industries. Additionally, cement production, which is responsible for around 8% of global CO_2 emissions, has high potential to damage the environment [7]. The building sector, which includes building maintenance, as well as the manufacturing, transportation, and processing of construction materials, is responsible for 38% of the world's CO_2 emissions [8]. If the Intergovernmental Panel on Climate Change's goal of limiting global warming is to be achieved [9], it is essential to reduce greenhouse gas emissions. There are various ways to achieve this goal at the level of building components, such as reducing resource consumption, extending service life and using recycled materials. The construction industry can play a significant role in achieving this goal, due to its high share of global greenhouse gas emissions.

Steel-reinforced concrete, a composite material made of concrete and reinforcing steel, has been the most important material in the construction sector since its invention in 1854 [10]. To reduce the required quantities of the materials, it is advisable to minimize the amount of concrete used. Textile concrete offers a way to reduce the amount of concrete needed for building components. Technical textiles replace steel reinforcements, resulting in lower energy consumption and CO_2 emissions during production [11].

Carbon fibers are an ideal material for textile reinforcements, due to their high tensile strength and corrosion resistance [11]. Carbon fibers have advantages as reinforcement material, but they also have some disadvantages. The latter include high costs, energy-intensive production processes and challenges related to end-of-life handling [12]. One approach to achieving the goals of the EU waste policy is to use recycled carbon fibers as an alternative reinforcement material for concrete. But the end-of-life treatment of carbon fiber is highly problematic as the fiber length of the recycled fibers is shortened, a fact which significantly limits further applications. Thus, recycled carbon fibers can be used in the construction industry as short-fiber-reinforcement or after the production of textile semi-finished products made of recycled carbon fibers from textile reinforcement [13]. However, compared with existing short-fiber-reinforcements made of steel fibers, alkali-resistant glass fibers, carbon fibers or PVA fibers [11], recycled carbon fiber does offer both advantages and disadvantages in fiber-reinforced concrete [13].

Recycled carbon-fiber short-fiber-reinforced concrete benefits from the corrosion resistance of the carbon fiber. Additionally, recycled carbon fiber has a lower price than virgin carbon fiber and still a high tensile strength. The recycling process employed determines the mechanical properties of the recycled fibers. Thermal recycling can recover fibers with 80–95% of the original fiber tensile strength [14]. Chemical recycling can achieve up to 98% of the original fiber tensile strength [14]. However, the low density of recycled carbon fiber causes the fibers to float to the top of the concrete and the small fiber diameter makes it more difficult to work with than with the distribution of macro fibers. Furthermore, the low elongation at break can cause brittle failure [13]. The suggested fiber volume content for recycled carbon fiber ranges from 0.5% to 1.5% [13]. This fiber content can lead to an improvement in flexural tensile strength by 11% to 40%, with a maximum of 10.7 N/mm^2, compared with steel and AR-glass-fiber-reinforced concrete. In addition, recycled carbon-fiber short-fiber-reinforcement can reduce primary energy requirements by up to 54%, depending on the application [13].

Figure 6.4 Application variants of recycled carbon fibers

Recycled carbon fibers have previously been used in the construction industry as fillers or substitutes for less efficient fiber materials [9]. However, these applications cannot be considered as efficient, value-preserving recycling, due to their lower requirements. To achieve value-preserving recycling, it is necessary to fully utilize the mechanical properties of the recycled carbon fibers. Therefore, there is interest in processing recycled carbon fibers into textiles, in addition to reusing them as individual short fibers. This approach offers the advantage of positioning the carbon fibers where they can provide the greatest mechanical benefit.

Nonwoven reinforcements (see Figure 6.4 (c)) are a potential application for recycled carbon fibers as textiles. However, when nonwoven reinforcement is used, the reinforcing layer may act as a two-dimensional separating layer in the concrete [13]. Besides its use as concrete reinforcement in Buildtech applications, recycled carbon fiber nonwoven lends itself to other reinforcement structures in carbon-fiber-reinforced plastic applications, for example, in the Mobiltech or Sportech categories. In use-cases with carbon-fiber-reinforced plastic, the separation layer in concrete is no longer a problem.

For Buildtech approaches, to produce a textile with a mesh opening, as is commonly done when virgin carbon fiber serves as the concrete reinforcement, it is necessary to produce yarns from recycled carbon fibers (see Figure 6.4 (d)). Hybrid yarns are produced by combining recycled carbon fibers with PA6 fiber material, as it is not yet possible to produce yarns from recycled carbon fiber alone. These hybrid yarns could be used as reinforcement for fabric production.

In the context of the circular economy, it is crucial to consider the production of components, including the extraction and production of materials, as well as the end-of-life phase, which includes the scope for recycling and the use of the recycled materials instead of new ones. The separability and therefore the recycling of textile concrete is significantly influenced by the material, the construction and the crushing method [15]. Separability can be enhanced by a suitable coating. An epoxy coating has given the highest recovery rates. Recycling is also positively affected by a larger roving cross-section. The hammer mill has been found to be the most effective method for separating carbon and alkali-resistant glass reinforcements, with a textile recovery rate of over 90% and a residual organic fiber content in the minerals of less than 0.2% [13]. As previously stated, recycling textile reinforced concrete as short fibers in fiber concrete is feasible. However, immediate reuse of the textile structure is not possible, due to resulting strength losses. The mineral content of the recycled textile reinforced concrete can serve as recycled concrete admixture in the production of concrete, roads and paths.

Initial investigations into textile waste as a reinforcement for concrete are underway. Textile waste from the manufacturing process in the form of fiber, yarn and textile has been used in these investigations, as well as end-of-life textiles from discarded clothing. The findings indicate that recycled textile waste fiber, especially nonwoven, is a technically feasible, sustainable and durable reinforcement for cementitious mortars used in low to medium performance applications [16].

6.1.4 Conclusion and Outlook

Technical textiles are used in diverse applications across multiple sectors, including industry, home, construction, agriculture, sports, transportation, healthcare, geotechnics, protection, packaging, clothing, and environmental protection. These textiles serve crucial roles in enhancing functionality and safety through innovations, such as thermal protection, structural reinforcement, smart textiles for health monitoring and sportswear, and sustainable solutions that use recycled materials. The integration of technical textiles into everyday products underscores their importance in modern industry and lifestyle improvements.

The importance of the circular economy in the field of technical textiles is growing, with the focus on extending the life cycle of materials through various "R-strategies", such as reuse, repurpose, and recycle. With sustainability concerns and resource scarcity on the rise, examples are provided to illustrate these strategies across different categories of technical textiles. Consideration of the entire life cycle of materials — from production to end of life within the circular economy framework with a view to enhancing sustainability is important. Due to the demands placed on technical textiles, it is not always possible to reuse them in the same area after their initial life cycle. However, there are several approaches that seek to reduce, repurpose and recycle the respective materials in different ways in the future, as described above. Nevertheless, significant future opportunities remain for enhancing the circular economy of technical textiles.

References for Section 6.1

[1]   Messe Frankfurt Exhibition GmbH. Profil der Techtextil. Wolfgang Marzin (Vorsitzender), Detlef Braun, Uwe Behm (accessed December 22, 2023).

[2]   Grand View Research. *Market Analysis Report. Technical Textile Market Size, Share & Trends Analysis Report By Manufacturing (3D Weaving, Thermo-forming, 3D Knitting), By End-use (Agro Textiles, Hometech Textiles), By Region, And Segment Forecasts, 2023 – 2030*; Report ID: 978-1-68038-154-2.

[3]   Fortune Business Insights Pvt. Ltd. *Markt Forschung Bericht. Marktgröße, Anteil und Branchenanalyse für technische textilien, nach Produkttyp (Agrotech, Buildtech, Clothtech, Geotech, Hometech, Indutech, Medtech, Mobiltech, Packtech, Protech, Sporttech und Oekotech), nach Fasertyp 8Naturfaser und Kunstfaser), nach Produktform (Stoff, Fasern und Garn) und regionale prognose, 2019-2026* Bericht-ID: FBI102716: India, 2020.

[4]   United Nations Environment Programme. Circularity: Circular economy processes. https://www.unep.org/circularity (accessed December 22, 2023).

[5]   Gagg, C. R. Cement and concrete as an engineering material: An historic appraisal and case study analysis. *Engineering Failure Analysis [Online]* 2014, *40*, 114–140.

[6]   Monteiro, P. J. M.; Miller, S. A.; Horvath, A. Towards sustainable concrete. *Nature materials [Online]* 2017, *16* (7), 698–699.

[7]   Ellis, L. D., Badel, A. F., Chiang, M. L., Park, R. J.-Y., Chiang, Y.-M. Toward electrochemical synthesis of cement-An electrolyzer-based process for decarbonating CaCO3 while producing useful gas streams. *Proceedings of the National Academy of Sciences of the United States of America [Online]* 2020, *117* (23), 12584–12591.

[8]   United Nations Environment Programme. *0 Global Status Report for Buildings and Construction: Towards a Zero-emission, Efficient and Resilient Buildings and Construction Sector*: Nairobi, 2020.

[9]   IPCC. *Global Warming of 1.5 °C;* Cambridge University Press, 2022.

[10] Scheerer, S., Proske, U. *Stahlbeton for Beginners. Grundlagen für die Bemessung und Konstruktion,* 2. Aufl. 2008; Springer-Lehrbuch; Springer Berlin Heidelberg: Berlin, Heidelberg, 2008.

[11] Friese, D., Scheurer, M., Hahn, L., Gries, T., Cherif, C. Textile reinforcement structures for concrete construction applications––a review. *Journal of Composite Materials [Online]* 2022, *56* (26), 4041–4064.

[12] Zhang, J., Chevali, V. S., Wang, H., Wang, C.-H. Current status of carbon fiber and carbon fiber-composites recycling. *Composites Part B: Engineering [Online]* 2020, *193*, 108053.

[13] Kimm, M. K. Ressourceneffizientes und recyclinggerechtes Design von Faserverbundwerkstoffen im Bauwesen. Dissertation; Rheinisch-Westfälische Technische Hochschule Aachen; Shaker Verlag, 2020.

[14] Gharde, S., Kandasubramanian, B. Mechanothermal and chemical recycling methodologies for the Fiber Reinforced Plastic (FRP). *Environmental Technology & Innovation [Online]* 2019, *14*, 100311.

[15] Kimm, M., Gerstein, N., Schmitz, P., Simons, M., Gries, T. On the separation and recycling behavior of textile reinforced concrete: an experimental study. *Mater Struct* 2018, *51* (5). DOI: 10.1617/s11527-018-1249-1.

[16] Payam, S. Sustainability, Durability, and Mechanical Characterization of a New Recycled Textile-Reinforced Strain-Hardening Cementitious Composite for Building Applications (Doctoral Thesis); Unpublished, 2022.

6.2 Composites Recycling

Mesut Cetin, Technische Hochschule Augsburg, Felix Teichmann and Georg Stegschuster, Institut für Textiltechnik Augsburg GmbH, Germany

Fiber-reinforced parts (FRP) are high-strength and extremely lightweight materials consisting of reinforcement fibers and a matrix system. The matrix system has the task of protecting the fibers, transferring the forces to the fibers and ensuring the shape of the component. The reinforcement fibers give the fiber-composite material the necessary strength. Figure 6.5 shows the structure of composites schematically.

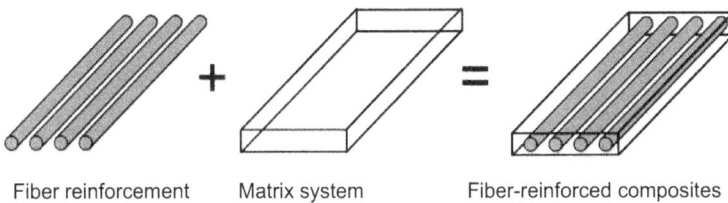

Fiber reinforcement   Matrix system   Fiber-reinforced composites

Figure 6.5 Structure of composites

Basically, all composite materials are structured in accordance with Figure 6.5. Glass, carbon and aramid fibers are the most widely used reinforcing fibers which are embedded in a thermoset (e.g., epoxy resin) or thermoplastic (e.g., polyamide) matrix [1–3]. Fiber-composite materials can be classified and defined on the basis of fiber type and matrix system. The use of carbon or glass fiber in combination with thermoset or thermoplastic systems dominates the market share, accounting for almost 70% [4], which is why the focus here is on these material combinations.

According to the latest figures from JEC (*https://www.jeccomposites.com/press/press-kit-jec-observer-overview-of-the-global-composites-market-2022-2027/*), the global volume of the composites market in 2022 totaled 12.7 million tons, which represents growth of about 5% over the 2021 figure of 12.1 million tons. The share of carbon fiber-reinforced composites in the total volume in 2022 was only about 1%, at approximately 0.1 million tons per year [5]. The European market has approximately 22% of the global composites market, as does America, while Asia accounts for the highest share, at about 50% [4].

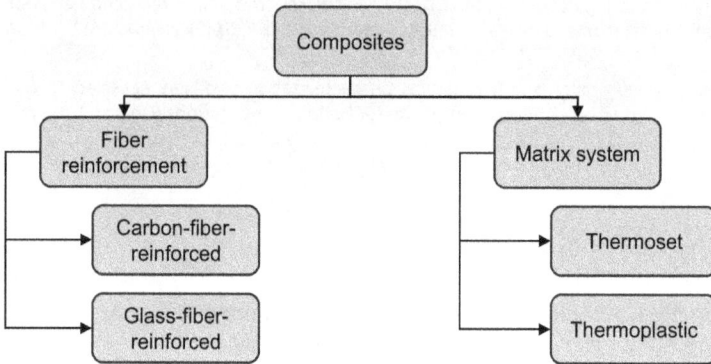

Figure 6.6 Most widely used composite materials

In terms of applications, the transportation sector, comprising the passenger car segment, commercial vehicles and aviation, accounts for the largest share at 51.1%. The electronic sector covers a significant portion at approximately 20.4%, with applications for switches, housings and control cabinets. The construction/infrastructure sector includes pipelines, containers, tanks and profiles (Figure 6.7) [4].

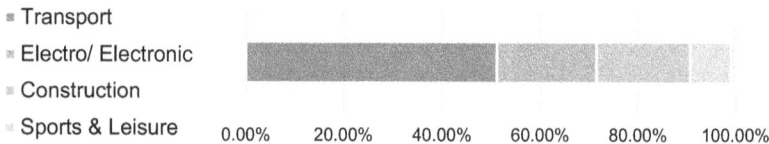

Figure 6.7 Share of applications in the composites market 2022 [4]

The global use of technical fibers in fiber-composite applications has steadily increased in recent years. Like many other sectors, this growth was dampened by the COVID-19 crisis. However, the market managed to recover. Compared with other traditional materials, such as steel or concrete, fiber-composite materials can reduce total weight while maintaining or improving the material properties of aircraft components, vehicle components, or transportation components. This weight reduction increases fuel efficiency. For electric vehicles, a lighter vehicle with the same battery capacity can travel a greater distance, as less energy is needed to move the weight. Additionally, reducing the carbon footprint of vehicles or products during use provides a significant advantage, due to weight reduction, and helps mitigate climate warming over the life cycle.

A typical application for FRP is hydrogen pressure vessels. "Type 4" hydrogen pressure vessels consist of a plastic inner liner (PA, PE) and a casing made of carbon fiber. This construction gives the pressure vessel an additional weight advantage over other designs ("Type 1–3") and the ability to withstand higher pressures. This leads to higher-capacity vessels. The applications for "Type 4" pressure vessels are in both the transportation and mobility sectors.

The use of FRP is one approach for achieving climate goals and resource savings. And, in this context, fiber-reinforced materials are gaining in importance. However, there are also challenges. The production of carbon fiber is a very energy-intensive process, which accordingly has a negative impact on the carbon footprint of the product. With the increasing use of FRP, the need for waste treatment is also growing and will be required by future regulations. This applies, for example, to products in the aerospace industry (life cycle: 20–30 years), the automotive industry (10–15 years) [6] and wind turbines (approx. 10 years). The replacement or rebuilding of wind turbines is driven by the elimination of subsidies and/or by replacement with more powerful wind-turbine units.

According to a study by [7], some 325,000–430,000 tons of glass fiber-reinforced components waste from wind turbine rotor blades and 76,000–212,000 tons of waste from carbon-fibre-containing rotor blades is expected to be generated in Germany by 2040. A similar scenario can be sketched for hydrogen pressure vessels. As soon as the use of hydrogen becomes established, high amounts of waste can be expected in later years. To ensure the use of composite materials while considering recycling, recycling processes must be developed to close the life cycle loop and reuse the valuable fibers.

6.2.1 Approaches to Composites Recycling

Recycling technologies for composites are presented below. According to the recycling pyramid, the following recycling approaches are only to be applied when the product cannot be reused otherwise. The recycling processes can be categorized as mechanical, thermal and chemical.

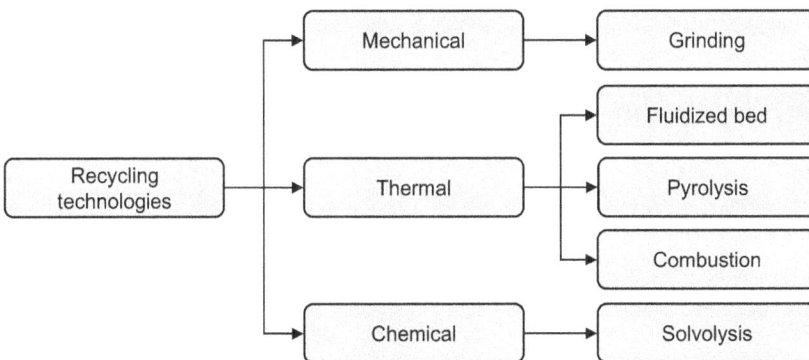

Figure 6.8 Common recycling technologies for carbon- or glass fiber-reinforced materials

6.2.1.1 Mechanical Recycling

Mechanical recycling refers to the process of recycling materials by mechanical methods. In the context of plastics, for example, mechanical recycling involves the collection, sorting, and shredding of components or products to produce new products or use them as fillers in other applications. The sorted FRP is mechanically shredded into small particles or chips. This can be done by cutting, grinding or shredding, the choice depending heavily on the application of the product and whether it involves thermoset or thermoplastic matrix systems. According to [8], mechanical recycling can be divided into two steps:

1. Destruction of existing material connections with the help of mechanical stress ("decomposition crushing").

2. Production of specific piece sizes ("piece size distributions") and shapes that are required and optimal for subsequent sorting.

For FRP with thermoplastic matrix systems, mechanical shredding and reprocessing into fiber-reinforced plastic pellets for reuse in injection molding technology are suitable both economically and process-wise. In the case of known and homogeneous amounts of waste (fiber-volume content), especially in manufacturing, these can be partially reintroduced into the process (cradle to gate). An example of this, an FRTP door module (PP GF 30, glass fiber-reinforced polypropylene with a fiber content of 30%) Figure 6.9, was investigated by [9].

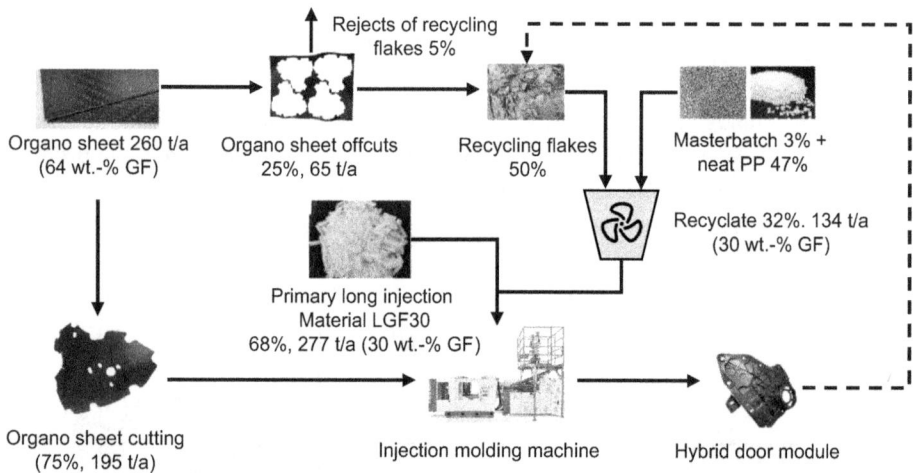

Figure 6.9 Recycling TFRP (cradle to gate) for the example of a door-module carrier [9]

It has been shown that recycled offcuts are comparable to virgin material in terms of mechanical properties and usability and they also have economical and ecological advantages. The recycling of glass fiber-reinforcement polypropylene in a closed loop is an effective way to reduce industrial waste in a sustainable and economical production process. [10] showed that the maximum tensile strength of samples (PA66 GF 35) does not behave linearly with the recycling content and that the average fiber length has an influence on the fatigue strength.

For known compositions (cradle to gate), adding reinforcing fibers or plastic material can achieve the target fiber-volume content and produce consistent components. However, for unknown waste compositions, reusing becomes more challenging, as the fiber volume content and the composition of the plastic may not be known. Such waste needs to be analyzed before it can be further recycled.

For FRP with thermosetting resin systems, the goal is the creation of a fibrous fraction which can serve as reinforcing fibers, for example, in sheet molding compound (SMC) and bulk molding compound (BMC).

6.2.1.2 Thermal Recycling

Thermal recycling can be classed as part of material recovery, because the objective is to reuse the raw material after the recycling process. Since the processes is energy-intensive, the associated costs are significantly higher than for mechanical recycling. This is the main reason why thermal methods are used for carbon fiber-reinforced components, as they have a much higher value than glass fibers. Thermal recycling of glass fiber-reinforced components is not economical. Methods employed in industry include conventional pyrolysis, microwave pyrolysis, and fluidized beds.

6.2.1.2.1 Conventional Pyrolysis

As a thermal recycling process, pyrolysis consists in thermal decomposition of the matrix system under exclusion of oxygen (inert atmosphere) by breaking the chemically highly cross-linked atomic bonds of the resin system. The process can be divided into five steps (Figure 6.10). Pre-sorting (1) is followed by the main pyrolysis process (2). Subsequently, the fibers can be sized (3) and then cut to a specific fiber length (4) and packaged (5). After successful decomposition of the resin system, only the fiber-reinforcement remains. The choice of process temperature depends on the matrix system and ranges according to [11] from 370–460 °C for polyester and epoxy resins and from between 450–580 °C for phenolic resins. In addition to recycled fibers (rF), other products are also obtained as a result. Gases and pyrolysis oils can be extracted during pyrolysis, directly reused, or further processed into other products. The combustible gases are often directly reused for the pyrolysis process to reduce energy consumption.

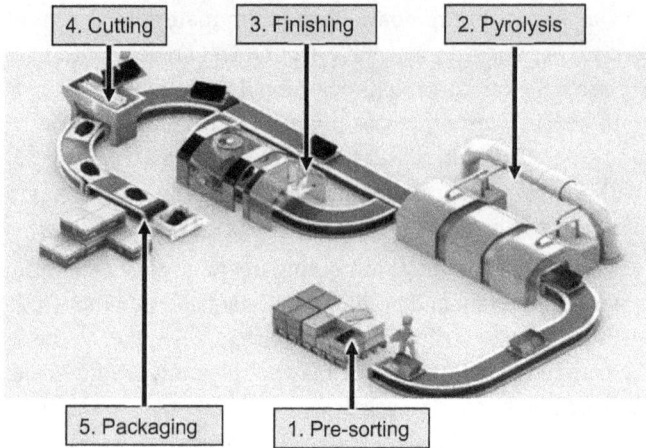

Figure 6.10 Example of a pyrolyzed carbon-fiber bicycle frame [12]

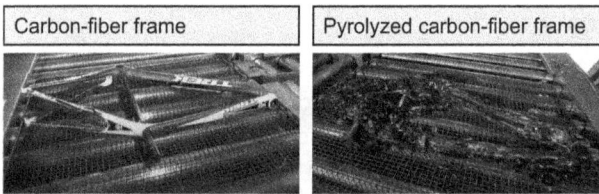

Figure 6.11 Carbon-fiber frame, before and after pyrolysis [12]

6.2.1.2.2 Microwave Pyrolysis

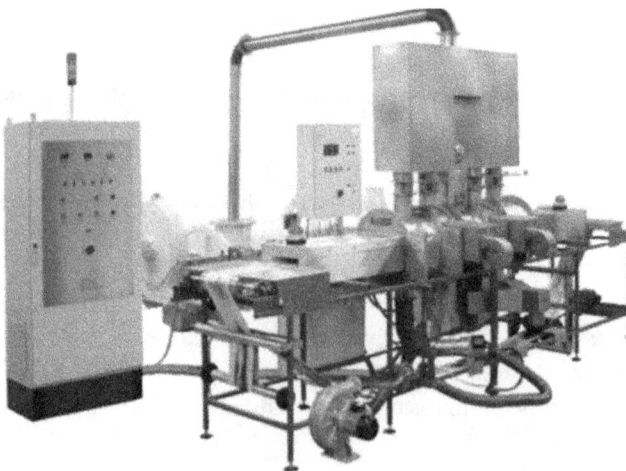

Figure 6.12 Microwave pyrolysis system (Linn High Therm GmbH)

Decomposition of the matrix system by microwave pyrolysis consists in using radiation to first heat the matrix by absorption and then decompose it [13]. In glass-fiber-reinforced plastics, only the matrix absorbs the radiation. In carbon-fiber-reinforced plastics, both the carbon fiber and the matrix absorb the radiation [7]. As with conventional pyrolysis, the output from the pyrolysis is the fiber-reinforcement, pyrolysis oils and gases. A significant difference from conventional pyrolysis is the process control. Achieving homogeneous heating with microwave radiation is challenging from a process perspective, as "hot spots" during may occur during application and can cause potentially fiber damage. Due to the complicated process, the recycling volume is also considered to be small (Figure 6.12).

6.2.1.2.3 Fluidized Bed

In contrast to the conventional pyrolysis, the fluidized bed process take place in an oxygen atmosphere at temperatures above 450 °C and at ambient pressure. In this process, FRP waste is shredded first and then added to the fluidized bed. The solid-gas fluidized bed is formed by the granular solid, through which hot gas flows. As the airflow increases, the sand at the vortex point acquires the characteristics of a liquid. This expanded, fluidized bed creates an intense mix of air and solid vertically and horizontally, leading to optimal heat transfer between the hot air and the solid materials (sand and composites). This generates high reaction rates in the fluidized bed and results in combustion of the matrix system. In the fluidized sand bed, the waste floats like a liquid, becomes uniformly distributed in the bed, and burns in the bed. The remaining lighter fibers are expelled from the reactor, separated and captured [8].

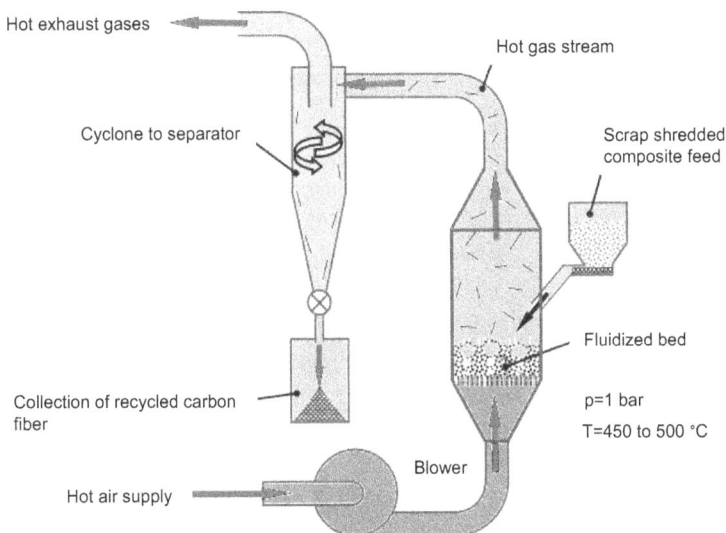

Figure 6.13 Fluidized bed process for composites recycling [14, 15]

6.2.1.3 Chemical Recycling

Solvolysis is a chemical recycling technology, by means of which the matrix system is decomposed by a solvent. The selection of the solvent, as well as the choice of the process values, is crucially dependent on the chemical composition of the matrix system, and it is necessary to prevent a chemical reaction on the fiber material by the solvent. Depending on the type of the matrix system, this reaction may take the form of hydrolysis (solvent water), alcoholysis (alcohols), or glycolysis (glycol). Other solvents are acids, alkalis, or ammonia. Solvolysis offers an opportunity to recover not only the recycled fibers, but also the decomposition products of the matrix system, which can be separated from the solvent in subsequent steps and prepared, for example, via polymerization [16]. The most important influencing parameters of solvolysis, in addition to the solvent, are reactor type, reaction time, temperature, pressure, the presence of a catalyst, and the process pressure [17]. To an extent depending on the temperature and pressure, solvolysis is classified as low-temperature ($< 200\,°C$), high-temperature ($> 200\,°C$), and supercritical (Figure 6.14).

The dissolution of the matrix system in the solvent is most effective under high pressure, high temperatures, and in supercritical fluids. However, under these conditions, the reactors must be of a rugged design and made with high-strength materials. Compared with pyrolysis, solvolysis is more environmentally friendly in terms of environmental impact. Neither CO_2 nor greenhouse gases are produced. However, depending on the application, solvent residues may remain.

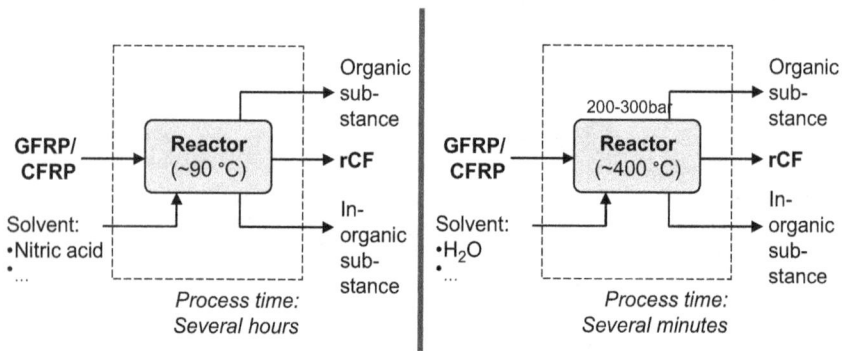

Figure 6.14 Solvolysis types for composites recycling

6.2.2 Evaluation of Fiber-Recycling Approaches

All the aforementioned methods lead to different ways of recycling of the fibers, which can be further processed into semi-finished textiles and used in new products. However, when fibers are reused, the fiber properties may not be fully returned to their virgin state. The damage to the fibers depends on the recycling technology and their specific process parameters.

In summary, a comparison of Young's modulus and the tensile strength of carbon fibers after recycling to the reference (before recycling and untreated) for the above-mentioned methods shows that the influences on the fiber properties are highly dependent on the specific process parameters of the recycling technologies (Table 6.1). Studies have shown that conventional pyrolysis has almost no effect on Young's modulus, but the tensile strength decreases by approximately 35% [18]. Microwave pyrolysis has a significantly greater impact. Here, a slight reduction in Young's modulus (approx. 9%) and a very strong reduction in tensile strength (approx. 75%) were observed [19]. A reduction in tensile strength by approximately 54% was also observed in the fluidized bed process, while Young's modulus remained almost constant [20, 21]. Solvolysis had hardly any influence on the key properties. Also, Young's modulus and tensile strength remain largely unchanged after solvolysis [22].

Table 6.1 Comparison of Young's Modulus, Tensile Strength between Virgin Fibers (Reference: 100%) and Recycled Fibers and TRL Evaluation

	Conventional Pyrolysis [18]*	Microwave Pyrolysis [19]**	Fluidized Bed Pyrolysis [20, 21]***	Solvolysis [22]****
Young's modulus [GPa]	≈ 100%	≈ 91%	≈ 100%	≈ 97%
Tensile strength [GPa]	≈ 65%	≈ 25%	≈ 46%	≈ 100%
Technical readiness level [-]	8–9	4–5	5–6	5–6
Capacity [kt/a]	< 5	< 0.05	< 0.1	< 1

*: Carbon fiber, 500–550 °C, N2
**: Carbon fiber, 45 GHz, 0.9 kW, 500–550 °C, 5 min.
***: Carbon fiber, Toray TORAYCA T700, 550 °C
****: Carbon fiber, water and acetone (80:20), 320 °C und 170 bar

Not all recycling processes are highly industrialized and have the potential to process large amounts of FRP waste. To obtain fibers lengths over 50 mm, only pyrolysis and solvolysis can be considered. The fluidized bed process requires a pre-shredding (3–6 mm) of FRP waste for efficient processing. The challenging process control and management in microwave pyrolysis leads to fiber damage, as temperature control is limited. Due to the process and technological advancement (TRL 4–5), its processing volume is also limited.

Despite the very gentle processing by solvolysis, the process is not yet fully industrially developed. The high demands on the machine technology and the discontinuous process make it difficult to process large quantities and the chemical process is limiting, which is the reason for the TRL of 5–6. However, companies and research institutes are working on increasing processing capacities.

Conventional pyrolysis is currently characterized by its high technical readiness level. It enables large quantities to be processed, thanks to the continuous process. In comparison with solvolysis, the fiber is damaged to a certain extent, which is due to the process itself. High temperatures and long reaction times quickly lead to damage to the carbon fiber through oxidation at a high temperature level, which is why process control is very important. Figure 6.15 clearly shows that temperature control has a major influence on fiber damage. Higher process temperature (a) leads to damage on the carbon fiber surface, which is also reflected in the fiber characteristics (Table 6.1). Oxidation on the fiber surface leads to break points and reduces the tensile strength. Also a change in the cross-section and diameter has been observed [23]. Compared with pyrolysis, solvolysis is gentler here, as it only dissolves the matrix, due to the solvent, but does not damage the fiber (c), which is also reflected in the characteristic values (Table 6.1).

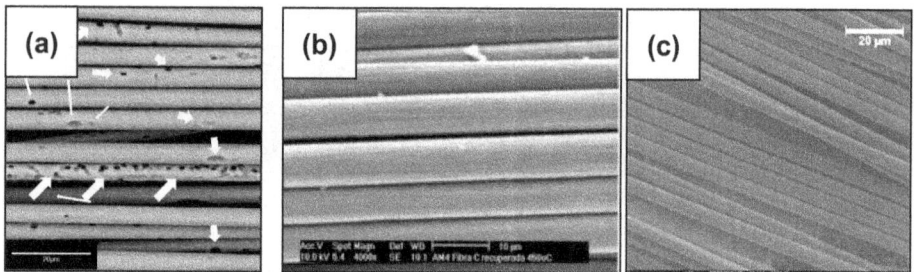

Figure 6.15 Scanning electron micrographs of surface of carbon fibers after conventional pyrolysis for 2 h at (a) 700 °C [24], (b) 450 °C [25] and solvolysis for 30 minutes at 400 °C and 25 MPa

6.2.3 rCF Textile Manufacturing

The recycling approaches for the fibers involve cutting, tearing, and shredding, which lead to a reduction in fiber length and an unknown length distribution. Additionally, the recycled fibers are present in a disordered and chaotic distribution. Knowledge of fiber length and fiber orientation is crucial for their use in fiber-reinforced composites. On one hand, the mechanical properties of the composite material increase with the average fiber length [26]. On the other, anisotropic carbon fibers achieve their maximum mechanical potential when subjected to stress in the fiber's longitudinal direction [3, 27].

Additionally, there are two types of waste, namely pre-consumer (textile cuttings, pre-forming cuttings) and post-consumer (recycled from component waste). Pre-consumer waste is also labeled as "recycled" and can be purchased commercially, containing 100% identical and sized fibers from one manufacturer with constant quantities. Thus, constant qualities and quantities are expected. Post-consumer waste, on the other hand, is more challenging, due to several unknowns, such as:

- Origin of the fibers in a batch (fiber type and manufacturer)
- Sizing of the fibers may no longer be present, due to the recycling process
- Fibers may be present in different forms
- Length of the fibers may vary from batch to batch

On account of these different aspects, the processing of post-consumer fibers is more challenging. Institutes and companies are using different technologies in different applications to tackle this subject.

The nonwovens fabric manufacturing (dry or wet nonwovens) process is particularly suitable for processing recycled fibers into a semi-finished textile, as it directly produces a semi-finished textile with uniform properties in terms of fiber distribution and alignment from the short, disordered fibers. Other manufacturing methods, such as weaving, laying, knitting, and weaving, that rely on continuous fibers and/or yarns [2] are limited, because of the finite fiber length (< 150 mm) and the unknown fiber length distribution

The use of recycled carbon fibers (rCFs) in new FRP components requires the rCF to be processed into a semi-finished textile product. Taking the technical readiness level of nonwoven lines into account, the nonwoven fabric manufacturing technology is characterized by cost-effective processing of various fiber materials with high production efficiency [28]. The baseline process for several semi-finished textiles is the carding process, followed by manufacturing processes for a woven, a tape, a nonwoven or a non-crimp-fiber textile structure (Figure 6.16).

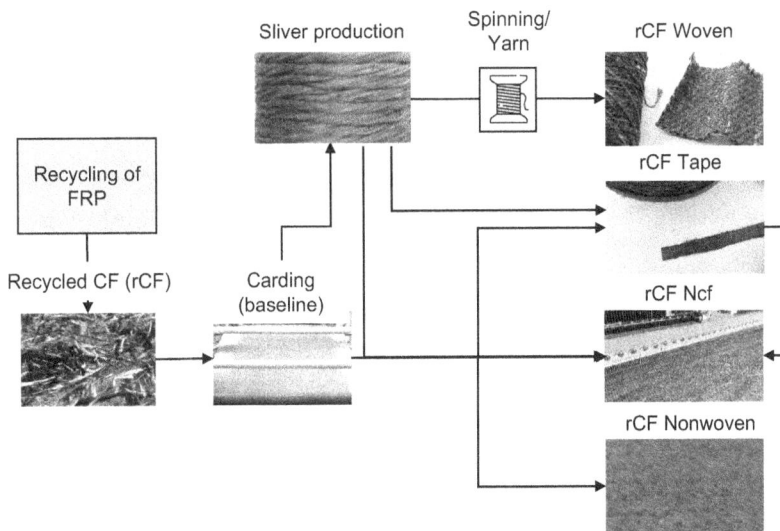

Figure 6.16 Overview of approaches for processing recycled fibers to semi-finished textile structures

Of the various nonwoven fabric manufacturing processes, carding is the most essential for producing semi-finished textile products, as it homogenizes and orients the unoriented recycled fibers and prepares them in a suitable form for further processing. Carding additionally has the advantage of being able to mix and process different kind of fibers (input material: glass, carbon or thermoplastic fibers) into any combination of materials. This enables the production of hybrid nonwovens, nonwovens made of 100% rCF or blends with natural fibers (Figure 6.17), which can then be converted into a FRP component in a thermoforming process or RTM process, for example.

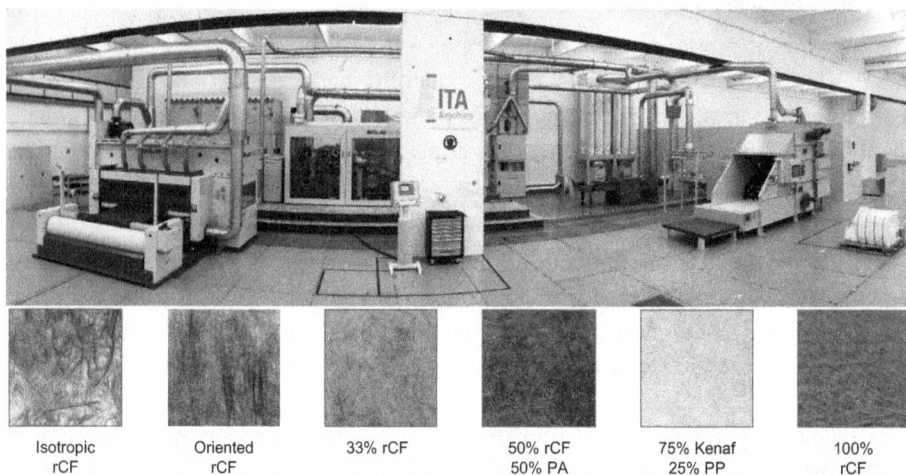

| Isotropic rCF | Oriented rCF | 33% rCF | 50% rCF 50% PA | 75% Kenaf 25% PP | 100% rCF |

Figure 6.17 Industrial carding nonwovens manufacturing line (DILO) at the Institute for Textile Technology in Augsburg (top) and some material samples (bottom)

Nonwovens are the most widely researched and most common semi-finished textiles. Approaches aimed at greater fiber orientation to meet higher mechanical requirements require other processing methods.

One approach is the processing of carbon fibers into a yarn, which is possible with the use of conventional textile machinery, to end up in a woven structure. Most challenging is the production of yarns from recycled fibers. The research project "CarboYarn" investigated the production of yarns made of recycled carbon fibers. All yarn types Figure 6.18 examined (combination of rCF material and spinning process) a strong reduction in rCF length during processing. No clear correlation could be shown between the different yarn types with the average rCF length and the spinning process. It could be shown that the pyrolyzed rCF (ELG) showed a significantly higher shortening in all investigated spinning processes compare to the investigated chopped fibers.

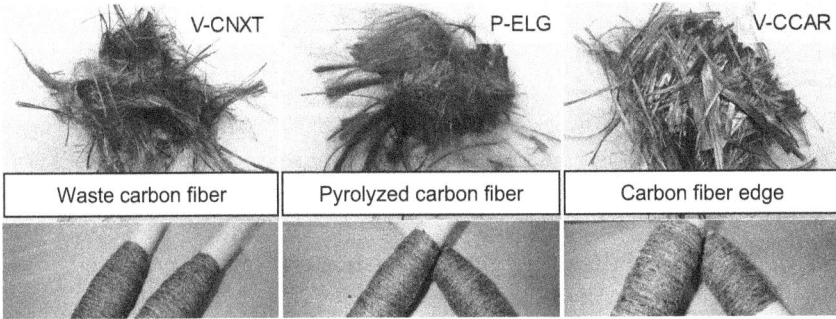

Figure 6.18 Processing of carbon fiber to a hybrid rCF yarn (V-CNXT: CarboNXT Chopped 60.000 NP5; P-ELG: C. Cramer T700; V-CCAR: ELG Carbiso C 40/100) [29]

The processes and rCF hybrid yarns are already within the economic range for pre-industrial or small-scale industrial applications. Costs and revenues are expected to be in the usual range for new fibers. In a cost comparison, ring yarn and covered yarns were similar. In fiber-composite applications, higher revenues were achievable with rewinding yarns, due to the greater orientation of the rCF in the yarn itself, but process improvements are still needed. Ring yarns are limited here by their structure and can be used as high-priced rCF yarns of low titer [29].

Another way to obtain a textile structure having a target carbon fiber orientation is the use of slivers instead of rovings for on a non-crimp-fiber machine (Figure 6.18). The main advantage over a rCF nonwoven is the possibility of achieving different oriented layups with greater fiber orientation. This process is still at the fundamental research level and has not yet reached industrial maturity. The challenge in this process relates to the preparation of the opened slivers, and provision and fixing in the machine.

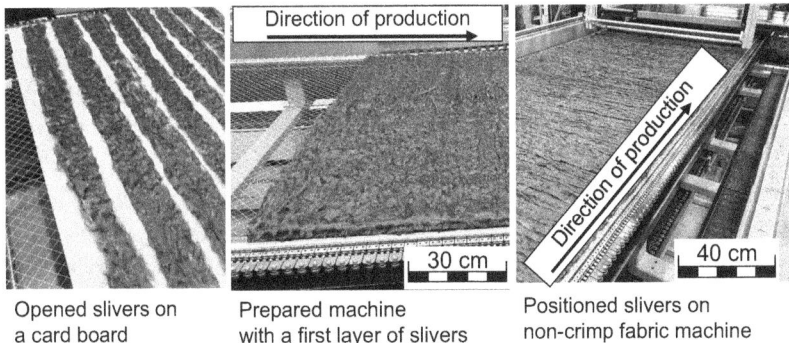

Figure 6.19 Processing of carbon fiber to an oriented rCF non-crimp fiber [30]

Another approach involves the production of recycled carbon fiber tapes (rCF tapes). In this method, recycled fibers are processed into a web using a nonwoven carding unit. The web is then shaped into a sliver, stretched, and formed into a tape. To stabilize the material in its tape shape, it undergoes consolidation. This is achieved by mixing thermoplastic fibers or other binder materials with the recycled carbon fibers. The overall goal for rCF tapes is to achieve a high degree of fiber orientation in the direction of the tape, which results in a high fiber volume content of the FRP parts. Fiber alignment or stretching can be used to increase the fiber orientation and can be simplified by two points. The first point feeds/clamps the fibers, while the second point pulls the fibers. This process achieves fiber alignment. However, if the alignment force inside the fiber is too high, there is a high risk of fiber damage or breaking. This can occur when the fiber is too long or is already fully aligned (between the two points). The extent to which a fiber can be aligned depends on several factors, including the clamping distance, fiber length, fiber-fiber friction, relative speed difference between the two points and applied force by the clamping points [26].

There are generally numerous variants of the downstream stretching technology/process, all of which are challenged by the short fiber length and the smooth surface of the rCF. The most promising research approaches trying to overcome these problems add more thermoplastic material. Stabilization of the stretching processes and therefore a high degree of orientation has been observed, with tensile strength and stiffness showing ratios of up to 17.3:1 and 12.2:1, respectively. For the bending strength and stiffness, ratios of up to 5.3:1 and 10.0:1 were determined, while an FVC of up to 45% was achieved [28].

6.2.4 Conclusion

To summarize, possibilities for manufacturing semi-finished textile made of rCF exist. But none yields a high degree of fiber orientation or a high output or has reached a high maturity level. rCF nonwoven textiles are established and on the market, but only made of pre-consumer waste materials. Post-consumer rCF nonwoven materials have so far failed to succeed, because of a lack of quality standards. Due to the process itself, nonwoven textile structures do not have a high degree pf fiber orientation. Thus, the applications are limited. To achieve higher fiber orientations for rCF, many research institutes are working on rCF tapes with a view to processing them on tape-laying machines, for example. The goal is a highly industrialized process, but technological improvements are still needed (TRL: 5–6). Nonetheless, the feasibility of producing oriented recycled carbon fiber semi-finished products (rCF tapes) on a scalable level has been successfully demonstrated. To leverage the high fiber orientation of rCF tapes for manufacturing large FRP structures similar to common non-crimp fiber textiles, rCF tapes are processed like rovings or tows in a non-crimp fiber machine. However, due to limitations of the rCF tapes and the process, the TRL and the output of this approach are low (5–6). Also, some research projects have explored developing rCF yarns and processing them into woven textile structures. Thanks to the industrialized weaving process, the TRL level of thise method is higher (7–8). Nonetheless, processing of rCF yarns like normal carbon fiber rovings or tows remains challenging, leading to insufficient output.

Table 6.2 Comparison of rCF Processing

Type of Product	TRL	Fiber Orientation	Output
rCF Woven	7–8	++	0
rCF Tape	5–6	+++	0
rCF Non-crimp fabric	3–4	++	0
rCF Nonwoven	9	+	+++

0 (not sufficient) + (sufficient) ++ (good) +++ (very good)

References for Section 6.2

[1] Witten, E., Mathes, V., Sauer, M., Kühnler, M., Composites Market Report 2018: Market developments, trends, outlooks and challenges (2018).

[2] Gries, T., Veit, D., Wulfhorst, B., Textile Fertigungsverfahren: Eine Einführung, 3rd Edition, Hanser, München, 2019.

[3] Schürmann, H., Konstruieren mit Faser-Kunststoff-Verbunden: Mit 39 Tabellen, 2nd Edition, Springer, Berlin, Heidelberg, New York, NY, 2007.

[4] Witten, E., Mathes, V., Der europäische Markt für Faserverstärkte Kunststoffe / Composites 2022.

[5] Sauer, M., Schüppel, D., Market Report 2022: The Global Market For Carbon Fibers and Carbon Composites – Market Developments, Trends, Forecast and Challenges.

[6] Kümmeth, M., Gottlieb, A., Ramerth, J., Seitz, M., Hartleitner, B., Rommel, W., Entwicklungsstudie zur Errichtung einer CFK-Recyclinganlage in Bayern: Entwicklung eines geeigneten Recyclingverfahrens am Beispiel der MPA Burgau.

[7] Kühne, C., Stapf, D., Baumann, W., Mülhopt, S., Entwicklung von Rückbau- und Recyclingstandards für Rotorblätter (2022).

[8] Martens, H., Goldmann, D., Recyclingtechnik, Springer Fachmedien Wiesbaden, Wiesbaden, 2016.

[9] Hummel, S., Obermeier, K., Zier, K., Krommes, S., Schemme, M., Karlinger, P., Analysis of Mechanical Properties Related to Fiber Length of Closed-Loop-Recycled Offcuts of a Thermoplastic Fiber Composites (Organo Sheets). Materials (Basel) 15 (2022).

[10] Bernasconi, A., Davoli, P., Rossin, D., Armanni, C., Effect of reprocessing on the fatigue strength of a fiberglass reinforced polyamide. Composites Part A: Applied Science and Manufacturing 38 (2007), 710–718.

[11] Blazsó, M., Pyrolysis for recycling waste composites. In Management, Recycling and Reuse of Waste Composites Elsevier, 2010, pp. 102–121.

[12] Hofmann, M., Gulich, B., Verarbeitung von rezyklierten Carbonfasern für die Herstellung von Verbundbauteilen. Lightweight Des 6 (2013), 20–23.

[13] Akesson, D., Foltynowicz, Z., Christeen, J., Skrifvars, M., Microwave pyrolysis as a method of recycling glass fiber from used blades of wind turbines. Journal of Reinforced Plastics and Composites 31 (2012), 1136–1142.

[14] Pickering, S. J., Recycling Thermoset Composite Materials, 1–17.

[15] Pickering, S. J., Kelly, R. M., Kennerley, J. R., Rudd, C. D., Fenwick, N. J., A fluidised-bed process for the recovery of glass fibers from scrap thermoset composites. Composites Science and Technology (2000), 509–523.

[16] Dorigato, A., Recycling of thermosetting composites for wind blade application. Advanced Industrial and Engineering Polymer Research 4 (2021), 116–132.

[17] Morin, C., Loppinet-Serani, A., Cansell, F., Aymonier, C., Near- and supercritical solvolysis of carbon fiber-reinforced polymers (CFRPs) for recycling carbon fibers as a valuable resource: State of the art. The Journal of Supercritical Fluids 66 (2012), 232–240.

[18] Kümmerth, M., Gottlieb, A., Ramerth, J., Entwicklungsstudie zur Errichtung einer CFK-Recyclinganlage in Bayern: Entwicklung eines geeigneten Recyclingverfahrens am Beispiel der MPA Burgau (29.02.2012).

[19] Emmerich, R., Kuppinger, J., Kohlenstofffasern wiedergewinnen. Kunststoffe 2014.

[20] Meng, F., McKechnie, J., Turner, T. A., Pickering, S. J., Energy and environmental assessment and reuse of fluidised bed recycled carbon fibers. Composites Part A: Applied Science and Manufacturing 100 (2017), 206–214.

[21] Pickering, S. J., Turner, T., Meng, F., Morris, C. N., Heil, J. P., Wong, K. H., Melendi-Espina, S., Developments in the fluidised bed process for fiber recovery from thermoset composites. 2nd Annual Composites and Advanced Materials (26.-29.10.2015).

[22] Jiang, J., Deng, G., Chen, X., Gao, X., Guo, Q., Xu, C., Zhou, L., On the successful chemical recycling of carbon fiber/epoxy resin composites under the mild condition. Composites Science and Technology 151 (2017), 243–251.

[23] Thomason, J., The Influence of Fiber Cross Section Shape and Fiber Surface Roughness on Composite Micromechanics. Micro 3 (2023), 353–368.

[24] Abdou, T. R., Botelho Junior, A. B., Espinosa, D. C. R., Tenório, J. A. S., Recycling of polymeric composites from industrial waste by pyrolysis: Deep evaluation for carbon fibers reuse (2021).

[25] Matielli Rodrigues, G. G., Faulstich de Paiva, J. M., Braga do Carmo, J. M., Botaro, V. R., Recycling of carbon fibers inserted in composite of DGEBA epoxy matrix by thermal degradation. Polymer Degradation and Stability 109 (2014), 50–58.

[26] Thomas, G., Thermoplastische Formmassen: Grundlagen, Verarbeitung, Anwendungen. 4. Aufl., Wiesbaden: Springer Vieweg (2014), 278–290.

[27] Stegschuster, G., Analyse des Kardierverfahrens zur Herstellung von Carbonfaservliesstoff als Verstärkungstextil für Faserverbundwerkstoffe, Shaker Verlag, Düren, 2021.

[28] Gulich, B., Faservliese nach aerodynamischen Verfahren. Wiley-VCH (2012), 158–171.

[29] Lechthaler, L., Stegschuster, G., Bell, E., Janssen, A., Gries, T., Schlichter, S., Ausheyks, L., Reichert, O., Baz, S., Gresser, G. T., Carboyarn: technologischer Vergleich von Spinnverfahren zur Herstellung von rCF-Stapelfasergarnen. Textil plus: die Fachzeitschrift für die textile Kette im deutschsprachigen Europa 7 (2019), 6–9.

[30] Emmerich, R., Dimassi, A., Huber, P., Uthemann, C., Dietrich, J., Teichmann, T., Özcelik, S., Abel, P., Gries, T., NEW CONCEPTS TO REDUCE THE ENVIRONMENTAL IMPACT OF FLOOR PANELS IN CIVIL AIRCRAFTS. ICCM 2023 – Recycling and sustainability (02.08.2023).

Index